Creating Level Pull

A lean production-system improvement guide for production-control, operations, and engineering professionals

by Art Smalley

Foreword by Jim Womack, Dan Jones, John Shook, and Jose Ferro

Lean Enterprise Institute
Cambridge, MA, USA
lean.org

Version 1.0
March 2009

With gratitude to Ron Sacco, Bryan Shipway, John Shook, George Taninecz, Helen Zak, and Design Continuum for their roles in the development of this workbook.

© Copyright 2004 Lean Enterprise Institute, Inc. All rights reserved.
Lean Enterprise Institute and the leaper image are registered trademarks of the Lean Enterprise Institute, Inc.

ISBN 978-0-9743225-0-6

Lean Enterprise Institute, Inc.
215 First Street, Suite 300
Cambridge, MA 02142 USA
(t) 617-871-2900 • (f) 617-871-2999 • lean.org

Acknowledgments

In preparing this workbook, I've realized my debt to the many fine supervisors and sensei who guided my education during my early career with Toyota Motor Corp. in Japan. Without their patient instruction and generous sharing of their extraordinary knowledge, I would not be able to pass along these lean concepts.

Additionally, I'd like to thank my good friends and mentors Tom Harada, Russ Scaffede, and John Shook for creating learning opportunities for me over the past 15 years. Special thanks are in order to my lean-expert friends and colleagues Elisa Martinez and Raoul Dubeauclard. They volunteered significant personal time and energy to review early versions of the workbook and supplied critical feedback as I crafted the final product.

Most importantly, I'd like to thank my wife Miwa for her loving support and cooperation while I wrestled with this project over many months. I promise to clean up the mess on the dining room table now that the workbook is finished.

Foreword

When we launched *Learning to See* in the summer of 1998 as the first publication of the Lean Enterprise Institute (LEI), we urged readers to start down a path toward perfect operational processes by mapping the value stream for each product family within the walls of their facilities. Our objective was to raise the consciousness of many managers from point improvements at the process level—creating cells, reducing set-up times, implementing 5S, improving the process capability of individual steps—to improvements in the performance of the entire value stream. We called this progressing from *process kaizen* to *flow kaizen*.

In the years since the launch of *Learning to See*, we've introduced additional workbooks describing how to introduce truly continuous flow in cellular production activities (*Creating Continuous Flow*) and how to implement a lean materials-handling system that supports continuous flow (*Making Materials Flow*). We've also extended the mapping process for product families far beyond the walls of individual facilities to encompass entire value streams (*Seeing the Whole Value Stream*).

Now we are ready to move beyond the value stream for individual product families and take on production control for all of the product families within a facility. We call this the leap to *system kaizen* because it ties together the flow of all products through a facility by means of a lean production-control system. To do this, many facilities will need to convert from traditional Material Requirements Planning (MRP) systems that schedule each activity within a facility and push product ahead to the next activity. Others will need to move beyond simple pencil-and-paper schedules or homegrown pull systems that do not effectively control or level production. In either case, the critical need is to transition to a rigorous *pull system* where each production activity requests precisely the materials it needs from the previous activity and where demand from the customer is *leveled* at a pacemaker process to smooth production activities throughout the plant.

To help you make this leap, we have asked Art Smalley to share his years of lean implementation experience. Art was one of the first foreign nationals to be made a permanent employee of the Toyota Motor Corp. in Japan where he worked at Kamigo Engine Plant, Toyota's largest machining operation. In 1994, Art left Toyota to become director of lean production at Donnelly Corp., an American automotive supplier with more than a dozen plants worldwide. In 1999, Art moved to McKinsey & Co. where he was a subject-matter expert on lean manufacturing and manager of McKinsey's Production System Design Center. In the course of his duties over the past 20 years, he has advised hundreds of facilities worldwide across a diverse set of industries on how to take a lean leap. In mid-2003, Art left McKinsey to spend more time at home with family, to write educational material on lean manufacturing, and to work directly with firms attempting a lean transformation.

We have warned in each of our workbooks that the step we are describing is harder than the steps required in previous workbooks, and we must offer this caution again. A truly lean production-control system that rigorously controls production at every step and levels demand from the customer has proved a great challenge for most firms. As a result, we usually see that this is the last element attempted in a lean transformation. If this is true in your case, you are in luck. In *Creating Level Pull*, Art has provided all of the basic knowledge you will need to get started in creating a lean production system in your facilities. And he has carefully constructed the workbook to be easily used by firms already far along with process kaizen and flow kaizen.

On the other hand, if you are just starting your lean transformation you also are in luck. Veteran lean practitioners usually urge firms with sufficient process stability to start their lean transformation by introducing lean production control with leveled demand as a system kaizen before moving to flow kaizen and process kaizen. If you are in this situation, we hope you will summon the courage to take the leap. The benefits for your business will be enormous and all of the knowledge you need is summarized here.

Given the nature of your challenge—wherever you are starting—we are anxious to hear about your successes as well as your difficulties and to connect you with the Lean Community at **lean.org**. Please send your comments to **info@lean.org**.

Jim Womack, Dan Jones, John Shook, and Jose Ferro
Cambridge, MA, USA; Ross-on-Wye, Hereford, UK; Cambridge, MA, USA;
Sao Paulo, SP, Brazil.

Contents

Foreword

Introduction

Part 1: Getting Started

Part 2: Matching Production-System Capability to Demand

Part 3: Creating the Pacemaker

Part 4: Controlling Production Upstream

Part 5: Expanding the System

Part 6: Sustaining and Improving

Conclusion

About the Author

Appendix

References

Introduction

Continuous flow of materials and products in any production operation is a wonderful thing, and lean thinkers strive to create this condition wherever possible. The reality of manufacturing today and for many years to come, however, is that disconnected processes upstream will feed activities downstream. Additionally, many internal processes are currently batch-oriented and function as shared resources. The major challenge in this situation is for downstream processes to obtain precisely what they need when they need it, while making upstream activities as efficient as possible. This is where *leveled demand* and *pull production* are critical.

As I visit manufacturing operations around the world, I rarely see anything resembling level pull production. Instead, I observe progress in introducing continuous flow as well as local stability improvements at the individual process level by means of point kaizen (such as 5S, enhanced process capability, and set-up time reductions).

The reason for this is not mysterious: creating level pull production in an operation of any complexity is not easy. Even within Toyota it took 20 years of hard work and experiments, between 1953 and 1973, to establish the system companywide. A successful transformation requires the coordinated efforts of everyone in a facility looking at the needs of all the product-family value streams. This calls for *system kaizen* of material and information flow to support every value stream.

Fortunately, the basic methods needed for level pull production are well understood, having been developed by Toyota and its affiliated companies over many years. In addition, there is now a considerable base of experience in introducing these methods in firms outside of Toyota. The challenge therefore is to provide a simple recipe for introducing these concepts in your facilities. Based on my experience in converting facilities from push to level pull-based production, I've developed 12 questions you will need to answer to meet the challenge. Not every question will apply in every case, and you may need to deviate slightly as your situation dictates, but I'm confident that by addressing these questions every facility can improve its performance while moving operations to the next level of sustainable achievement.

Only you can supply the courage and leadership to create level pull in your facilities. And much additional point, flow, and system kaizen will be needed after your initial leap. But this workbook provides all of the necessary knowledge to get started and to move you past the critical threshold from erratic push to level pull. I'll be anxious to hear about your experiences, and I wish you smooth sailing on level seas!

Art Smalley
Huntington Beach, CA
April 2004

Part 1 Getting Started

Part 1 Getting Started

Part 1: Getting Started

Welcome to Apogee Mirror

Apogee Mirror is a typical discrete parts manufacturer, making exterior mirrors, interior mirrors, and door handles for the automotive industry. Several years ago, Apogee responded to pressure from its customers for lower prices, higher quality, more frequent deliveries exactly on time, and more rapid response to changing market demand by taking a hard look at its manufacturing operations.

Apogee managers took a value-stream walk to follow the manufacturing paths of its three main product families. They drew value-stream maps for each product family—one of which is illustrated below by the map for the exterior-mirror product family. They soon were able to see wastes of many sorts: long set-up times on molding machines; poor uptime in the paint booth; many disconnected operations for assembling the product; and long throughput times with large inventories between every step in the process.

Original-State Value-Stream Map for Exterior Mirrors

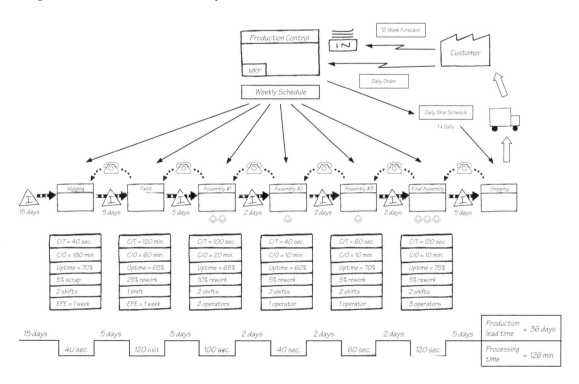

By energetically pursuing both point and flow kaizen along the three value streams, Apogee management and employees soon were able to achieve much better performance for all three product families, as shown by their current-state map for the exterior-mirror product family.

Current-State Value-Stream Map for Exterior Mirrors

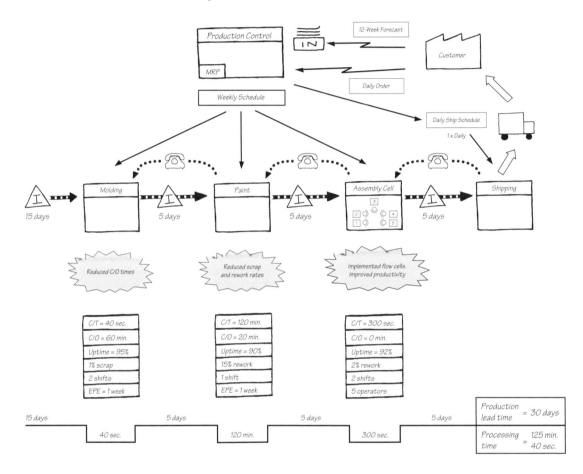

Apogee reduced changeover times in all of the processes, improved uptime in paint through point kaizen, and created compact continuous-flow assembly cells through flow kaizen. Because of this, Apogee managers were able to shrink throughput time and inventories while reducing effort and cost. They also were able to reduce the amount of manufacturing space required (*see Apogee Overhead Layout*).

Apogee Overhead Layout

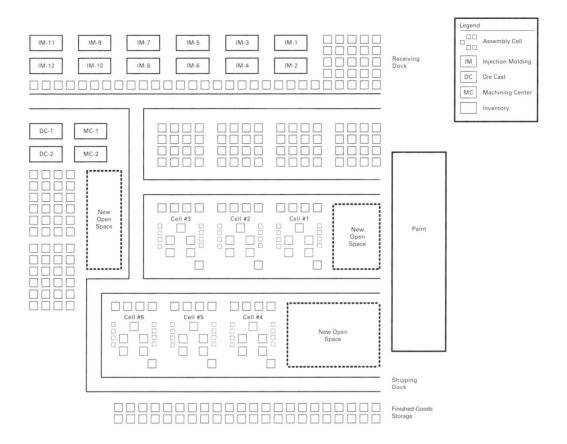

Like many companies today, Apogee avoided taking any action to more tightly link and control the flow of information between the production departments—molding, die casting, paint, assembly, and shipping. Apogee managers judged that modifying the information management system linking these areas—which pushes products ahead to the next processing step with the help of material handlers who respond as needed—would be complicated because the necessary systems would affect every value stream in the plant. In addition, many managers wondered if this leap really was necessary. They thought that sufficient improvements could be wrung from point kaizen and flow kaizen.

The Continuing Challenges of Delivery and Cost

Initially, Apogee's managers were delighted with their achievements as a result of point and flow kaizen. Morale in the facility was higher as a 5S program brightened work areas and employees participated in the kaizen activities. And direct-labor costs significantly declined.

Many dimensions of performance, however, did not improve as hoped. In particular, the facility still needed considerable overtime and expediting of shipments to meet customer requirements.

And while total inventories had been lowered, they still were high. Equally troubling, the reduction in direct-labor costs had not been matched by any change in indirect labor. Managers still were spending large amounts of time revising production schedules as customer requirements changed. Meanwhile, an army of material handlers raced through the plant to get the right materials to the right place to meet changing customer requirements.

Even more troubling, performance in some areas seemed to be deteriorating as the initial excitement of the kaizen initiative wore off. In particular, the paint, assembly, and shipping departments often reported that they could not provide what their customers wanted because of a lack of materials in the right place at the right time. This trend is shown in the box score.

Box Score—Exterior-Mirror Value Stream*

	Original state	After basic stability	After cell flow kaizen	Current state
Productivity				
Direct labor (pieces per person per hr.)	9.0	10.0	11.5	11.0
Material handlers supporting value stream	3	3	3	4
Quality				
Scrap	5%	4%	3%	2%
Rework**	25%	20%	15%	15%
External (ppm)	500	250	125	105
Downtime***				
Assembly (min. per shift)	40 min.	30 min.	10 min.	20 min.
Paint (min. per shift)	30 min.	20 min.	15 min.	15 min.
Molding (min. per shift)	50 min.	25 min.	25 min.	10 min.
Inventory turns				
Total	8	11	14	12
On-time delivery				
To assembly	65%	68%	80%	75%
To shipping	80%	92%	95%	85%
To customer	100%	100%	100%	100%
Door-to-door lead time				
Processing time (min.)	126.0	126.0	125.7	125.7
Production time (days)	36	34	28	30
Costs				
Overtime costs per week	$6,000	$5,000	$4,000	$5,000
Expedite costs per week	$2,000	$1,500	$1,500	$2,000

   * No major change in demand or product mix over this time.
  ** Rework is due to persistent inclusion problems.
 *** Downtime is separate from changeover time and reflects only lost time in production due to mechanical problems or material availability per shift.

The pattern of visible improvements at many points but limited progress in the facility as a whole, along with ominous backsliding in some improved areas, seemed to suggest that something was wrong with the total production system, not just the individual parts. Apogee managers therefore decided to take another walk to focus on the flow of information and materials between production areas and to look at the entire production system involving all three product families. What they saw was quite startling.

Traditional Scheduling in a 'Lean' Facility

The management team started its walk in the shipping area, following the value stream for exterior mirrors. It quickly learned from the Production Control manager that customer schedules were forecast well in advance and formed the basis for the weekly schedules sent to each production area by the computerized Material Requirements Planning (MRP) system. However, the weekly schedules bore only limited resemblance to the daily releases from customers that determined what was actually shipped. Because the throughput time in the plant from raw materials to finished goods was still several weeks, the frequent change in customer orders, as reflected in the daily releases, often meant:

- The wrong items—too many and too early—were being produced far upstream.
- Downstream processes, such as assembly, lacked the correct parts despite holding large inventories of many parts.
- Downstream processes had no effective mechanism to let upstream processes know what parts they needed next, short of supervisor intervention.

To deal with these problems, Production Control spent most of its time revising schedules and expediting parts within the plant. Yet during a normal shift only 75% of orders were ready to assemble on time, and only 85% were ready to ship on time. Because no automotive supplier can risk stopping its customer's assembly plant, Apogee dealt with the problem of products arriving late at the shipping dock by running large amounts of overtime every day (to get the product out of the plant that night) and by using expensive air freight. Senior managers also discovered on their walk that production capacity for each process was greater than the average demand. This meant that expensive overtime mostly was caused by scheduling problems rather than capacity constraints.

To illustrate the production-control problem Apogee was facing, the management team drew a simple graph (*see Apogee Demand Variation for Exterior Mirrors on page 6*) in which the variation in orders for exterior mirrors was plotted. The solid line shows the actual variation in weekly demand for units from the end customer over the most recent 13-week period for one of the two assembly cells in this value stream. (The two cells were identical. One produced right-side mirrors and the other produced left-side mirrors.)

Apogee Demand Variation for Exterior Mirrors

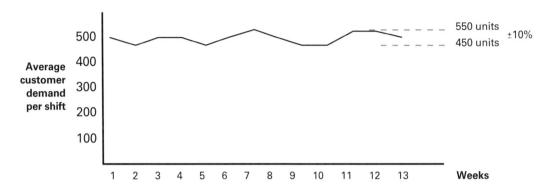

The chart reflects only variation in the total number of mirrors demanded from the left-side exterior-mirror cell. The situation became more interesting when mix variation for mirrors by colors and configurations was included for the same 13-week interval. A sampling of the top 10 part numbers that ran through the same assembly cell in the exterior-mirror value stream generated the chart below (*see Apogee Demand and Mix Variation*). The obvious conclusion was that total demand varied only slightly, but mix varied substantially.

Apogee Demand and Mix Variation

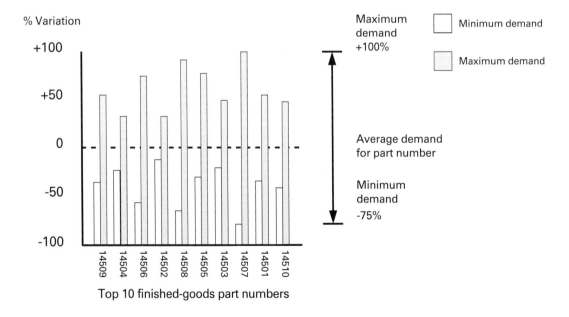

More surprising, however, was the realization that variation in both total demand and mix got progressively worse through the plant. By collecting data on actual production orders at each process step, the Apogee team soon was able to see that the variation in daily release amounts was less than the variation in actual orders sent to the two assembly cells for exterior mirrors, and this variation was less than that experienced by the most upstream production step for this product family (molding). In short, Apogee faced a modest challenge from erratic customer demand—one no worse than what most facilities face—but its internal practices made the problem much worse than it needed to be.

Demand Transmission and Amplification

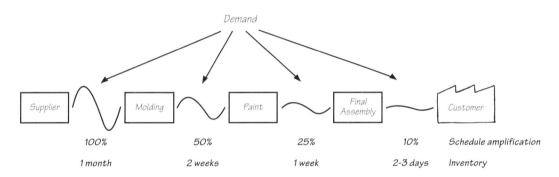

The Need to Switch from Erratic Push to Level Pull

As the Apogee managers reflected on their walk, they suddenly understood some simple ideas they had read about but never really grasped. They had discovered that their customers were only modestly erratic with their orders, but Apogee's internal scheduling practices were making the situation much worse by transmitting customer variation to every step in the production process and then increasing the variation. They also realized that the centralized production-control methods, which attempt to schedule every point in the production process, were pushing products ahead to the next production area on the basis of the forecast rather than the actual needs of the next process. This was causing inventories to pile up ahead of every step.

The team realized Apogee had a *cognitive* scheduling system that pulled all information up to a centralized point for decision making when they really needed a *reflexive* production-control system permitting each production point to signal its needs to the previous production operation. In biological terms, Apogee was transmitting all information to its brain for processing when it really needed to let its reflexes take over. When we put our finger on a hot stove, we don't methodically review the situation and propose the best course of action. Instead, our reflexes do the right thing by pulling our finger away. This is the simplest way to think about the difference between push and pull.

Similarly, the team could see that their facility was exposed to the full brunt of customer orders, as if built on an unprotected coastline and exposed to storm waves. Yet humans always have sought to locate important sea-related activities in safe harbors rather than on unprotected coasts. The purpose of the harbor and its breakwater is to prevent disruptive waves from reaching the docks, even though the level of water still rises and falls over time in the harbor to match the average level of the ocean. Apogee's location on an unprotected coast was allowing waves of customer demand and mix fluctuations to flood through the plant unchecked, becoming even more turbulent as they passed from department to department.

What Apogee needed was a way to level and smooth external customer orders to protect the activities within the plant from chaos while still serving the customer and while letting every production activity pull the materials it needed next from the previous process. They needed to *create level pull*!

Level and Pull to Smooth Demand Amplification

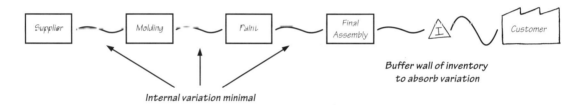

As we soon will learn, there are many places to locate strategic buffers that protect operations from demand waves while serving the customer better *and* improving operational performance. Indeed, much of this workbook will be devoted to locating and precisely sizing the appropriate buffers. The key point now grasped by the Apogee team was that inventories at the right points could greatly improve productivity and customer response.

> **Do You Have Sufficient Stability to Embrace Level Pull?**
>
> An important question for you to ask at this point is whether you have sufficient stability in your operations to move forward with a pull production-control system. In general, if individual processes have uptimes of 75-80%, as they did at Apogee, you can move forward on pull. However, if the output in many of your processes is less stable and predictable, lead times internally will vary tremendously and pull production will be very hard to implement. In these cases you probably will do better to spend a bit more time on point and flow kaizen to improve stability before attempting to make the leap to level pull.

At the end of their value-stream walk, Apogee managers resolved to install a truly lean production-control system for every value stream in their facility. This workbook will show how they did it, describing the questions they asked, the actions they took, the performance targets they set (*as shown in the chart below*), and the timeline adopted for the initiative.

Box Score—All Value Streams

	Original state	Current state	Target state
Productivity			
Direct labor (pieces per person per hr.)	7.8	10.2	12.5
Material handlers per shift	24	25	15
Quality*			
Scrap	5%	2%	<1%
Rework	25%	15%	<5%
External (ppm)	500	105	<50
Downtime			
Assembly (min. per shift)	40	30	<5
Paint (min. per shift)	30	15	<10
Molding (min. per shift)	20	20	<10
Inventory turns			
Total	8	10	30
On-time delivery			
To assembly	60%	75%	98%
To shipping	85%	85%	100%
To customer	100%	100%	100%
Door-to-door lead time			
Processing time (min.)	126.0	125.7	125.7
Production lead time (days)	36	30	12
Other			
Overtime costs per week	$30,000	$25,000	$0
Expedite costs per week	$12,000	$9,000	$0

* Quality issues will not be directly addressed in this implementation effort. These targets represent long-term goals for the value stream.

Implementation Approaches

I often encounter debates about which implementation path to take—narrow or broad, fast or slow—and I always say, "It depends." It depends specifically on:

- Your level of knowledge and experience as you start;
- The level of acceptance of the concept within your implementation team;
- Your need for quick results as opposed to the need to get it right the first time while educating a larger number of individuals;
- The nature of your production assets; and
- Your tolerance for making mistakes.

Firms with limited knowledge and experience, ambivalent managers in some key positions, and limited tolerance for errors will do better to follow the incremental path described in this workbook. Other firms with more knowledge (perhaps including an experienced external sensei), more buy-in, and lots of courage may follow the all-at-once path and will gain the full benefits of the transformation sooner. However, the end objective and the methods to employ are the same. With the information provided in this workbook, you can successfully follow either path or some path in between that best fits your circumstances.

As Apogee set out to create level pull, it needed much more than performance targets and the right questions. With a little experience you will find that setting reasonable targets is the easy part, and the questions to answer are always similar among facilities. *Apogee's most critical needs were the correct management team to spur the transition, a clear plan to guide everyone's actions, and a reasonable scope and timing for their efforts.*

The Transition Team

Apogee knew it was important for everyone in the facility to be involved, and created a special team for the transition. The team was led by a dedicated leader from Production Control, the organization that will operate the system over the long term. The team included one manager from every area of the operation—shipping, final assembly, paint, molding, die casting and machining, receiving, materials handling, industrial engineering, and human resources. The detailed implementation work then was done by a small staff, which worked full time on the project and reported frequently to the team.

Apogee set a six-month timetable to get the job done, which is reasonable in all but the largest facilities. The timetable listed every task to accomplish, established times to start and complete each task, and assigned clear responsibility for each task to a specific member of the team.

The Scope and Timing

Apogee could have started with the whole facility, done a lot of planning, and switched from erratic push to level pull on a given Monday morning. (And some facilities actually can do this. Yours may be one.) However, Apogee was attempting this conversion with no prior experience operating a level pull system. In addition, Apogee had limited resources—they could devote only a few full-time staff to the project. The area managers on the team needed to perform their normal tasks in their areas and could devote only a few hours each week.

Apogee decided to proceed in stages: They started with only one product family—exterior mirrors—and began their implementation at the shipping dock for this one product. They then worked backward to the two assembly cells for exterior mirrors, then to the paint booth, and finally to injection molding. At the end of the first two-month phase they had created a level pull system for only this product family. It was not very efficient from a total plant performance point of view because the rest of the facility, including a fraction of each shared process (paint and molding), was still operating on the old production-control system. But it worked and it demonstrated the concept. Based on their learning and the growing acceptance of the concept among formerly ambivalent managers, Apogee then transitioned the rest of the plant in a disciplined manner over four months.

Apogee faced a considerable challenge in tackling system kaizen to complement previous initiatives with point and flow kaizen. But the benefits were enormous. So, let's get started and follow their progress.

Part 2: Matching Production-System Capability to Demand

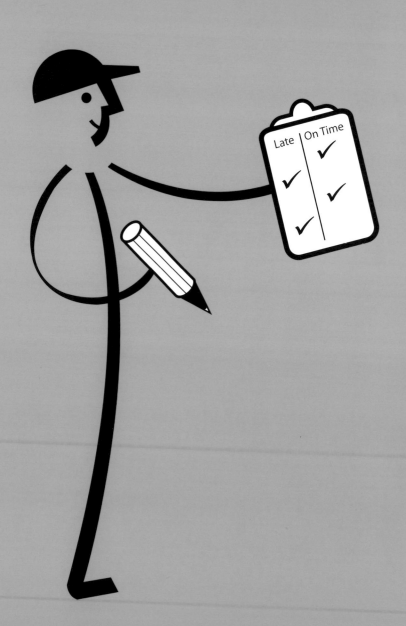

Part 2: Matching Production-System Capability to Demand

1. Which products should you hold in a finished-goods inventory, and which products should you produce only to a confirmed order?

2. How much of each product should you hold in finished goods?

3. How will you organize and control the finished-goods store?

Part 2: Matching Production-System Capability to Demand

To implement a level pull system, Apogee's improvement team needed to start with one product family—in this case, exterior mirrors—at the point closest to the customer. This meant thinking first about the point of shipment at the end of their internal value stream.

> **The Need for Simplicity**
>
> Two assembly cells were needed to meet the volume requirement for exterior mirrors, one cell for left-side mirrors, the other for right-side mirrors. This division simplified fixturing and speeded changeovers, and it also served customer requirements, which called for receiving lefts and rights separately.
>
> Apogee began its implementation with only one of the assembly cells, the one for left-side mirrors. This is the path we will follow until the end of Part 3, when we add the second cell.
>
> As you implement your own level pull system, it is very likely you will need to make similar simplifications. Remember that if you have limited experience or resources it is better to start with a simple path and make rapid progress than to tackle too much complexity and get bogged down.

Apogee always served its customer with timely shipments, which also was the price of entry in its industry because massive auto-assembly plants must never be stopped due to lack of parts from a supplier. Apogee tackled this problem by trying to maintain large buffers of finished goods for *every part number*.

When parts shortages emerged, despite the large stocks of finished goods, Apogee compensated first with overtime and then by expediting products through the production process. In either case, smooth operations upstream in the facility were disrupted. Because the existing system worked poorly and Apogee had new insights gained from the value-stream walk, the team asked three simple questions:

1. **Which products should Apogee hold in its finished-goods inventory and which products should instead be produced only upon receipt of a confirmed customer order?**
2. **How much of each product should Apogee hold in finished goods to protect both the customer and the facility from disruptions?**
3. **How should Apogee organize its finished-goods area to make management of the inventories easy?**

These are the first questions you will need to answer as you start to implement your own level pull system.

But Aren't All Inventories Waste?

Isn't inventory one of the seven wastes in lean manufacturing? Yes, Toyota does list inventory among the seven forms of waste.

So why should you maintain or even increase inventories of at least some finished goods? The reason is because not having a sufficient inventory of material can create even bigger wastes all the way up the value stream in the form of waiting, excess transport, excess effort (expediting) and overtime, and intermediate goods inventories.

While it would be ideal to have a 100% build-to-order system with practically no lead times, no production shortfalls, no capacity constraints for dealing with demand surges, and no finished-goods inventory, this is not practical for most industries with short customer-lead-time requirements. Inventory in the right place is a powerful tool to buffer against surges in external demand as well as against internal process instability. Until you develop the capability to build 100% on-time with practically no production lead time, and you have worked with your customer to smooth demand surges, you are likely to require finished-goods inventories for many items to act as a buffer.

Extra capacity and *longer response time* also can act as buffers, but both carry costs, either for you in the form of extra investment or for your customer in the form of wait time. Attempts to dramatically eliminate all inventory will meet with failure unless you first eliminate the problems that require you to hold inventory. Put another way, the popular saying, "lower the water level to see the rocks," sounds great, but all you will gain is a nice view of some rocks and a hole in your boat unless you have a way to remove the obstacles before proceeding. Above all, be practical in the initial stages of creating level pull in your operations.

Question 1:
Which products should you hold in a finished-goods inventory, and which products should you produce only to a confirmed order?

Like many manufacturers, Apogee had varying demand and mix for the different end products flowing through its exterior-mirror value stream. Twenty-five different left-side mirrors were made, varying by color and complexity (e.g., manually adjusted and power mirrors, heated and unheated mirrors, and color combinations). The chart below shows the distribution of demand for each part number as a fraction of total orders.

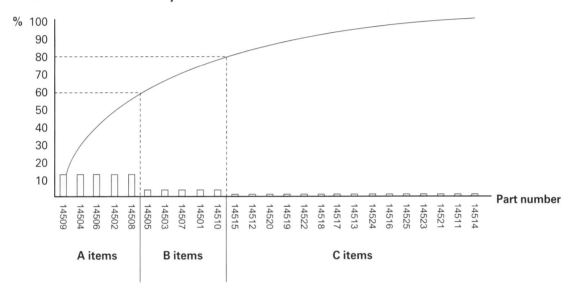

Distribution of Demand by Part Number

The bars in the diagram show the fraction of total demand accounted for by each part number. The curved line running from left to right stacks the orders to show the fraction of total demand accounted for by any given number of products. For example, the first five part numbers account for 60% of total demand and the first 10 account for 80%.

Apogee used these data to conduct a product segmentation that lean thinkers call an "ABC production analysis." (Do not confuse lean ABC analysis with a common practice at many companies of categorizing inventories as A, B, or C by annual dollar volumes.) In doing this they noted that five of the 25 part numbers accounted for 60% of demand and were ordered every day by the customer. These were the five *A items* or *high runners*. A second group of five products accounted for another 20% of demand and were ordered frequently but not every day. These were the five *B items* or *medium runners*. The third group of products consisted of the remaining 15 part numbers and also accounted for 20% of demand, but each was ordered infrequently and in highly variable amounts. This group—*C items* or *low runners*—included infrequent color and build combinations, special-edition items, and replacement parts, many of them for out-of-production vehicles. (Some manufacturers apply different names to their A, B, and C items, such as "runners," "repeaters," and "rogues," respectively.)

Apogee had been treating all three types of parts in the same way through the same centralized scheduling process and maintained considerable stocks of finished goods for every part number. However, due to the batch sizes in the plant and the lengthy lead time to produce parts, the plant still encountered out-of-stocks in finished goods, necessitating expedited production and shipment.

Deciding about Finished Goods vs. Make-to-Order

With their ABC production analysis in hand and knowledge about the ability of their production process to deliver on schedule, the Apogee team was ready to decide which products to hold in finished goods and which to make-to-order. They constructed the following chart as a way to list the logical options.

Options for Finished Goods vs. Make-to-Order

Options	Pros	Cons	Apogee situation	Apogee decision
1. Hold finished-goods inventory of all products (As, Bs, and Cs) and make all to stock—*replenishment pull system*.	Ready to ship all items on short notice	Requires inventory for each part number and much space	Finished-goods stores and shipping unable to hold all items	Not practical due to physical layout constraints and number of end items
2. Hold no finished-goods inventory and make all products to order—*sequential pull system*.	Less inventory and associated waste	Requires high process stability and short lead time to produce	Production lead time too long and paint process too unstable	Not practical with current lead time and capability
3a. Hold only Cs in inventory and make A and B products to order daily—*mixed pull system*.	Less inventory	Requires mixed production control and daily stability	Daily stability a concern	Possible second step for future
3b. Hold A and B products in finished-goods inventory. Make Cs to order from semifinished components—*mixed pull system*.	Moderate inventory	Requires mixed production control and visibility on C items	Most applicable to current situation	Best fit for today

Option #1 would be the safest and easiest to implement, but this required too much space (and too many inventory dollars) in finished goods to hold adequate amounts of every part number. In addition, Apogee needed to create more space in the facility for product launches in the upcoming year.

Replenishment Pull System

Holding finished goods for every type of product and using withdrawals by the customer to trigger production—as Option #1 requires—is known in lean manufacturing as a *replenishment pull system*: only consumption of the end items triggers replenishment of product. In this case the production instruction would be sent to the final assembly line from finished goods via the heijunka leveling device (to be explained shortly), and then back upstream from assembly.

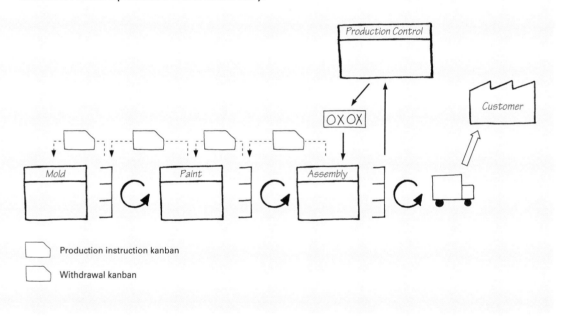

Option #2 was impractical at present because of the 30-day production lead time and process instabilities, particularly in the paint area. No customer was willing or capable of placing 100% firm orders so far in advance, and Apogee couldn't guarantee that all items could be made on time even if they did.

Sequential Pull System

Producing all items to customer order—as required by Option #2—is known as a *sequential pull system*: items are paced and built in accordance with demand, with the build instruction sent to the first process step at the beginning of the value stream. This type of pull system is more demanding to manage than a simple replenishment pull system because it is hard to pace the flow of operations to takt time. Unless your facility has a short and steady internal lead time and high equipment availability, this option will be difficult to maintain. Even in lean companies, such as Toyota, sequential pull is employed only when the operations have demonstrated high capability, and special build-to-order situations are required by the downstream process or customer.

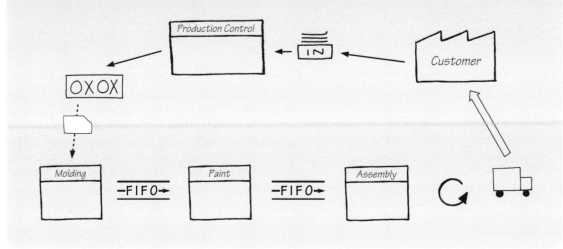

Option #3a was attractive because it minimized inventories of high-volume A and B items. But it required the assembly cells to receive good parts from paint in time to consistently deliver A items and B items on schedule to the finished-goods area. The team decided that this could be a goal for the future if production lead time could be sufficiently reduced and the capability of the paint process improved. But this option carried too much risk for the present situation.

Option #3b was chosen as the best fit between the needs of the customer and the current performance of the facility. This is a version of a mixed pull system where some items are built-to-order while others are replenished to stock. This option required holding appropriate finished-goods inventory for the A and B items, while assembling the 15 C items only to customer order. (Parts 3 and 4 will explain how the Apogee team handled the C items.)

Mixed Pull System

The scheduling required for Option #3 utilizes a *mixed pull system*: aspects of both sequential and replenishment pull are utilized in conjunction. This is particularly useful when the majority of items requested are frequent repeat orders, but many infrequent items are required as well. (Note that Apogee chose to assemble C items to order from parts held in the market after the paint process. The diagram below, by contrast, shows the more common situation in which orders for C items are sent to the beginning of the value stream at molding.)

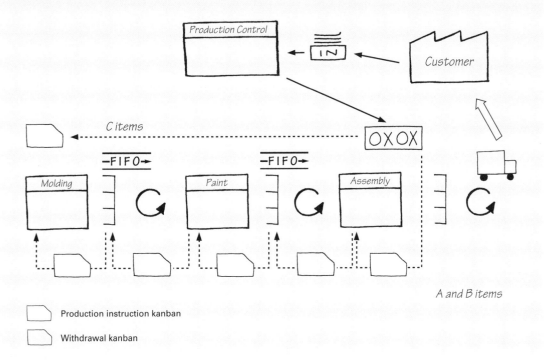

Choosing this option meant that Apogee spent less time managing the 80% of volume accounted for by A and B items, as long as they were consistently replenished on a daily or every-other-day basis. This enabled Apogee managers to concentrate daily efforts on the more management-intensive C items.

Handling A, B, and C Items: Together or Separately?

You may need to take a step back and think hard about whether C items should be produced in the same assembly cell as A and B items. If the work content between the items is quite different or the changeover pattern complex, you may want to consider putting C items in a highly flexible assembly cell dedicated to low-volume production runs. This C cell also could include low runners from other product families.

Having a dedicated cell for C items may simplify the pull system for the A and B items. Additionally, in industries where the profit margins on C items, such as spare parts, are much higher than for A and B items, it may make economic sense to produce them in separate cells even if the cost is a bit higher.

In Apogee's case, because the work content and the changeovers in assembly for C items are not disruptive, C items will be produced in the same assembly cell. In your own implementation you will need to think carefully about which option to follow. Either way, taking the time to weigh the options and collect the necessary data on current plant performance and customer demand patterns will pay large dividends in the performance of your level pull system.

Question 2:
How much of each product should you hold in finished goods?

Having decided to employ a mixed pull system and hold finished goods for the A and B items, the next question for the Apogee team was how much of each of these items to hold. After some discussion, the team adopted a simple formula to calculate the initial finished-goods inventory levels.

Finished-Goods Calculation

	Average daily demand x Lead time to replenish (days)	**Cycle stock**
+	Demand variation as % of Cycle stock	**Buffer stock**
+	Safety factor as % of (Cycle stock + Buffer stock)	**Safety stock**
=		**Finished-goods inventory**

Finished-Goods Calculation for Part #14509

	160 x 5*	Cycle stock	800
+	25%** of 800	Buffer stock	200
+	20%*** of (800 + 200)	Safety stock	200
=		Finished-goods inventory	1,200

   * Production Control schedules production of this part number once a week and average daily demand is 160.
  ** Reflects two standard deviations of demand and thus approximately 95% of normal order variation. If necessary, more standard deviations can be taken to cover a higher level of variation.
*** Reflects the worst-case example of scrap, rework, and typical downtime amounts at Apogee.

Because of the importance of this formula, it is worth walking through the calculation briefly, as Apogee applied it to part #14509:

The first step is to determine average demand per day. Apogee specified this using a three-month time span. (Of course, you can use a longer or shorter period depending on seasonality and likely changes in average demand due to market conditions.) This amount, 160 pieces for part #14509, is then multiplied by the number of days between scheduled replenishment (*the lead time to replenish*) for this part number—five days. Because the part is produced only every fifth working day, the maximum *cycle stock inventory* immediately after the part is produced must be five days worth or 800 pieces (160 x 5). Otherwise, supplies of this part number would be exhausted before the next scheduled production date. During the five-day cycle the amount on hand will fall steadily—it will cycle down and thus the term "cycle stock"—until it reaches zero just at the point of the next replenishment (*see Finished-Goods Inventory Example on page 22*).

What if there is a surge in demand during the five-day period before the next replenishment? The second factor in the calculation, *buffer stock*, deals with this problem by adopting the statistical technique of standard deviations. Standard deviations are used to calculate how likely it is that demand will surge by more than a certain amount during the replenishment period of five days. Apogee decided to set their buffer stock at a level of two standard deviations above average demand, which means the odds are 19 out of 20 that demand will not exceed the amount set in the buffer. (You, of course, can pick higher or lower odds.) Two standard deviations in demand equated to 25% above 800 pieces, or an additional 200 pieces. (Any introductory statistics text gives the formula for calculating standard deviations and there are simple functions in spreadsheets that will do this automatically for you as well.)

What if there is a shortfall in production from downtime or quality losses and the expected number of new parts doesn't arrive in finished goods when planned? The third and final term in the formula, *safety stock*, deals with this problem. Based on data collected from the previous three-month period, Apogee calculated that the maximum possible shortfall was 20% of the combined cycle stock and buffer stock (20% of 1,000), or 200 pieces. Apogee's paint department represented the least stable part of the production process, and had worst-case rework rates of up to 15% on any given day. Additionally, minor downtime and scrap losses totaled about another 5% in the plant. (You may choose a lower or higher percentage, of course, based on analysis of your production reliability and capability to respond to problems. Just be sure to rely on real data from the recent past rather than promises from production managers that they will be more reliable in the future. Remember, all inventory amounts can be adjusted downward over time as improvement occurs. Starting too high is much less of a problem than starting too low.)

By adding the maximum cycle stock (800), buffer stock (200), and safety stock (200), Apogee calculated that the maximum number of pieces of finished goods for part #14509 was 1,200 pieces. It also was easy to see that the minimum number of pieces—at the point just before the weekly replenishment—would be 400 pieces if everything was running normally.

Apogee selected this finished-goods inventory formula for two reasons: First, it was simple and everyone could understand it. Equally important, it emphasizes internal production lead time to replenish (frequency of replenishment), which highlights the importance of reducing lead time as the best internal way for operations to reduce inventory. This focus on production lead-time reduction will be critical to the success of the project as Apogee moves forward.

Finished-Goods Inventory Example

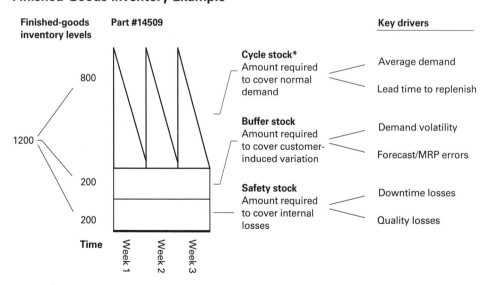

* Assumes an average customer demand and draw down during each of the three weeks. Greater than average demand or production shortfalls due to quality problems and equipment downtime will require Apogee to consume some buffer and/or safety stock.

After doing their math for all the required part numbers in the value stream, Apogee calculated an overall need for finished goods for the exterior-mirror value stream that was about the same as the total finished-goods inventory currently on hand—*but the inventory now was practically all A and B items.* In the future, the only C items in the finished-goods area will be products already completed as part of a large customer order.

To transition to the new system, Apogee's Production Control Department issued instructions to Operations to run overtime on the weekend to build up the necessary inventory levels for the A and B items. The existing C items were put into a special location and labeled carefully for use in the interim until they were made only to order.

Question 3:
How will you organize and control the finished-goods store?

Apogee's project team also decided to alter the way they stored finished goods. The plant had relied on a system that scanned the inventory as it entered the finished-goods area. The product then was placed in any open location on the shelves, and the material handlers scanned in the location as well with a handheld device. Although fine in theory, the system was prone to human errors and parts were frequently "lost," only to turn up later in an incorrect location. Also, first-in/first-out (FIFO) was hard to maintain because there was no visual queue as to which items to ship first. Apogee needed to combine the benefits of scanning with the simplicity of a visual method to manage the finished-goods inventory.

To do this, Apogee took to heart the lean concepts of visual control and workplace organization. The team decided to create dedicated locations for each part number, which were carefully designed to ensure that the oldest parts were shipped first. The improvement team also hung signage above the storage locations clearly indicating the maximum quantities for each finished item. This enabled the team to organize finished goods in a logical and visual fashion that allowed Apogee to help distinguish normal from abnormal circumstances. It also eliminated the daily search for products and improved the efficiency of the Shipping Department as they picked the daily order for the customer.

As a tracking and daily management tool, Apogee's Production Control Department created a spreadsheet for finished-goods inventory. Each part number was divided into the three categories: cycle stock, buffer stock, and safety stock. All three levels for A and B items were tracked and updated by the shipping department on a regular basis at the end of the first shift. The numbers constantly moved (especially in the cycle-stock category, as the customer took inventory away before replenishment at the end of the period), but it still was easy to see whether inventories were within the normal range. This tool helped Shipping and Production Control agree on the status of the inventory, what needed to be shipped next, and the high-priority items for production.

Tracking Spreadsheet for Exterior-Mirror Finished-Goods Inventory

Part #	Finished-goods inventory			Tues. 3:05 p.m.
	Safety maximum/actual*	**Buffer** maximum/actual*	**Cycle** maximum/actual*	
A items				
14509	200/200	200/200	800/900	Over by 100
14504	175/175	175/175	700/600	OK
14506	150/150	150/150	600/450	OK
14502	125/125	125/125	600/300	OK
14508	100/100	100/50	600/0	Into buffer stock
B items				
14505	100/100	100/100	300/100	OK
14503	100/100	100/100	200/50	OK
14507	100/100	100/100	200/50	OK
14501	100/100	100/100	200/100	OK
14510	100/100	100/100	100/0	Out of cycle stock
C items**				

* The actual amount is recorded at 3 p.m. each day after the first-shift daily shipment and completion of first-shift production.

** C items are built to order and not held in finished-goods inventory. Under the current temporary plan, C items are blended into production between runs of the A and B items, using overtime when necessary.

To simplify flow within the facility, all finished goods were stored after assembly near the shipping dock. The buffer and safety inventories were marked clearly on the holding racks and kept separate from the cycle stock. Control rules were adopted as well:

- Buffer inventory could be removed from finished goods by Shipping only with the authorization of the Production Control Department.

- Operations and Production Control needed to jointly review usage of buffer inventory to determine if it represented a fundamental change in customer demand or merely a one-time occurrence.

- The plant manager needed to authorize any use of the safety-stock inventory. Depletion of finished goods to the safety-stock level indicated a serious production problem requiring immediate countermeasures followed by investigation for the root cause.

When shipping releases were received, the Shipping Department picked the necessary pieces from the finished-goods storage location for the A and B items several hours in advance of every shipment. For the time being, Production Control manually scheduled the items that needed to be replenished by final assembly after the daily shipment, basing this decision upon calculations involving the depletion of the finished-goods inventory.

For the short term, Production Control also handed the production supervisor a list of the C items to produce, giving the supervisor the list several days before they were due. Using available production time between A and B items during the week and overtime when necessary, Operations made the C items. Shipping collects C items as completed in Operations and places them in the shipping lane along with the correct A items and B items for the next delivery truck.

Because the assembly cell was building batch sizes larger than the amounts normally taken away by the customer each day, Apogee could not yet build directly to dedicated lanes in Shipping and in small amounts. However, the plant was moving closer to building to daily replenishment for A and B items and closer to build-to-order for C items. (*Note: This does not represent the final Apogee solution. As Apogee progresses, we'll see how they improve the handling of daily production control and devise system improvements to refine the process, but only after a sufficient foundation is in place to move forward.*)

Make It Clear: Normal vs. Abnormal

If possible, organize and segment your inventory as Apogee has done. The main goal is to make clear to everyone whether inventory levels are *normal* or *abnormal*. Doing this correctly pushes down the ownership and control for inventory closer to the assembly cell and surfaces problems on a real-time basis for the managers and employees. Additionally, it can eliminate the need for extensive computer-generated reports, which typically are kept out of sight or entirely within the production-control system.

Matching Production-System Capability to Demand—Keys to Success

- Always go to the gemba (the production floor) to get facts, data, and anecdotes. Don't rely upon second-hand information.

- Determine if your production processes are stable enough to work around internal delays by buffering. If not, find the root cause and solve the problems before proceeding with your level pull system.

- Segment your demand by A, B, and C items, and determine what type of pull system works best for you.

- Carefully calculate the amount of cycle, buffer, and safety stocks for each finished-goods part number required to ensure on-time shipments to the customer without overtime or express freight.

- Practice good workplace organization and visual control in the finished-goods store and shipping area.

- Strive to make the status of finished goods—normal vs. abnormal—clear at all times to everyone.

Part 3 Creating the Pacemaker

Part 3 Creating the Pacemaker

4. Where will you schedule the value stream?
5. How will you level production at the pacemaker?
6. How will you convey demand to the pacemaker to create pull?

Part 3: Creating the Pacemaker

Apogee managers now had a firm grasp of their customer demand and the complications it was causing the plant. As a first step to stabilize the situation, they had determined which parts they would hold in finished-goods inventory, how many of each part to hold, and how to organize the finished-goods area for better visual control.

After one week, the results of their activities in finished goods were encouraging: the number of expedited orders during the week fell by about one-third and the amount of daily overtime attributable to expediting decreased from one hour per day to 40 minutes. (This was after the one-time buildup of A and B items required to change the composition of the finished-goods inventory.) Finished goods were ready to ship 92% of the time, up from 85%.

The Apogee improvement team was pleased to see that the symptoms had eased, but they knew they had not resolved the root cause of the problems. Moving material in a timely fashion between departments and into finished goods was still difficult because the assembly, paint, and molding processes were responding to multiple schedules.

To address the root cause of their inability to ship on time, the Apogee team needed to determine where to schedule the value stream, how to level production at this location, and how to convey demand to this location. Doing so would aid in controlling production throughout the plant.

Question 4:
Where will you schedule the value stream?

The team reviewed Apogee's value-stream map for the left-hand exterior-mirror value stream, this time focusing on the flow of information.

Left-Hand Exterior-Mirror Value Stream

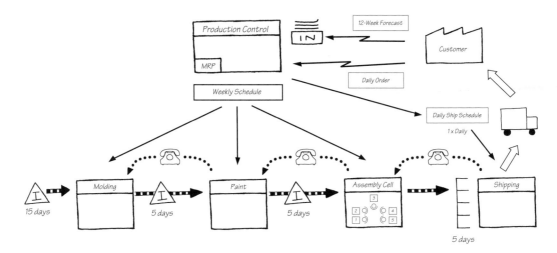

The map showed three distinct locations where scheduling occurred on a weekly basis; four locations if you counted the daily ship schedule. (Because molding and paint were shared with other value streams, their schedules were particularly complicated.) As indicated by the telephone calls from each process back to the previous process—sending instructions to ignore the schedule and produce the parts needed immediately by downstream processes—the more scheduling points in a value stream the greater the chance for errors.

From their earlier data collection, the team had gathered a number of anecdotes from the team leaders and operators on the shop floor about the actual use of the schedules by molding, paint, and assembly:

- "The schedule changes at least five times per week, so I don't put much faith in it at all."—*assembly cell operator*
- "I wait to be told what to make from paint; that way I know it is urgent."—*molding operator*
- "There are a few part numbers that really matter so I run them in the biggest batch possible until I have to change over."—*molding operator*
- "The schedule is more of a guideline; from experience we know what needs to be made."—*paint department supervisor*
- "I get measured more on productivity than schedule adherence, so I just try to make lots of parts."—*assembly cell operator*

As shown by these comments, scheduling information throughout Apogee quickly became outdated. And even if it was correct, it often was ignored. Multiple scheduling points combined with inherently unreliable forecasts, long lead times, large batch sizes, and department-centric metrics led the Apogee molding, paint, and assembly departments to a state of "What is best for me?" instead of "What does my customer need next?" The result was excess inventory of most parts combined with a failure to produce all of the right parts in the right amount at the right time for the customer. To overcome these problems, the Apogee team needed to designate a single point in the value stream as the *pacemaker* to receive the schedule from Production Control.

> **Traditional Scheduling Can't Keep Up, But Don't Throw It Out**
>
> Apogee's performance and the comments of production operators and first-line managers are typical of facilities that employ traditional scheduling techniques. In theory, sending multiple schedules from a central MRP system to each department should keep everyone informed and working to the same cadence. In reality it rarely does. Problems inevitably enter into the equation when assumptions for lead time, scrap and yield rates, and other inputs are wrong.
>
> The early-generation MRP programs have algorithms that assume infinite capacity in the system, a condition that never exists in the real world. They also frequently build in extra buffer or safety inventory at each step of the process. Recent systems are more refined, but the shop floor is still a dynamic place. It changes minute by minute throughout the day while MRP systems typically work with a time fence of anywhere from a shift to a week. The MRP needs to be continuously updated about the actual status of production on the floor, but this is difficult to achieve. Often schedules, production status, and inventory levels are only updated overnight in a batch program, making them useless for resolving problems arising throughout the day.
>
> Does this mean you should yank out your MRP system or develop an anti-IT bias? Certainly not. But you should recognize the inherent weakness and limitations of existing systems. Manufacturing companies always will need some type of MRP system to hold the bill of material, create rough-cut capacity plans, handle forecast information, and complete other useful tasks in production planning. However, even the most advanced software systems poorly execute *real-time, shop-floor control* for production between processes. This failure is what we have observed at Apogee.
>
> Pull production in a value stream as regulated by a pacemaker has great advantage over most standard software applications: With shop-floor production control, employees can sense and react more rapidly to changing production dynamics. There is no delay while waiting overnight or until the next schedule can be created and transmitted to the floor. Response to a problem can be nearly instantaneous. Additionally, shop-floor control puts responsibility and capability for solving problems in the hands of those actually running the process.
>
> Information technology continues to advance, but it remains a challenge for standard scheduling systems to incorporate the full logic of lean.

Guidelines for selection of the pacemaker process

Apogee needed to select one process to function as the overall pacemaker in the value stream. This would eliminate confusion over the "right" schedule, and it would ultimately enable everyone to march to the same beat—takt time at the pacemaker process. Coupling this beat with shop-floor control tools, especially kanban, would facilitate a smoother pull of material on time to the customer through the value stream.

Apogee kept the following in mind when selecting their pacemaker:

- In replenishment pull, final assembly will be the pacemaker in virtually every case.
- In sequential pull, the pacemaker often is the first process at the beginning of the value stream. If possible, however, the pacemaker should be located further downstream just before inventory types proliferate.

At Apogee, it was easy to see that the final assembly cell should be the pacemaker for the A and B items. The C items, however, presented a dilemma. One approach was to schedule the C items at the molding process, but the instability of the paint process between molding and assembly meant that the parts might not show up on time in assembly. (This often is a problem with sequential pull because you must maintain rigid FIFO control and pace the flow of parts to sync with takt time in final assembly. Otherwise you wind up with sequential push instead of pull!)

To simplify the problem given current stability, Apogee decided to create an inventory of painted parts for C items in a market between paint and final assembly, and to assemble these items only when an order was received. (To do this, Apogee will use a scheduling method that soon will be described.) This greatly simplified the mixed pull system by creating only one schedule point—in final assembly—for the value stream.

Question 5:
How will you level production at the pacemaker?

Perhaps without realizing it, Apogee had already taken the first step in leveling output at the pacemaker. This had happened two years earlier when Apogee created its continuous-flow cells working to takt time with standardized work. The operations team had balanced the two exterior-mirror cells and their daily output to a takt time of 54 seconds, meaning that 500 pieces were produced during every 450-minute shift. The previously erratic daily output had been replaced with output *leveled* in terms of the *quantity* produced per shift. (Of course, when customer demand soared for several days or plummeted briefly to a much lower level, the cells either worked some overtime or stopped early. The key breakthrough was getting the cells to produce at a level rate whenever operating and to stick with that rate until takt time was changed when long-term demand changed.)

Now Apogee needed to level production by mix by reducing the batch sizes produced of each part number in the cells to better reflect the amounts actually requested by customers in their daily shipping releases. A quick look at the chart (*see Apogee Orders vs. Apogee Build*) showed that Apogee's current batch size of 500 pieces for every item was far out of sync with typical daily requirements.

Apogee Orders vs. Apogee Build Schedule

Part #	Demand category	Mirror description	Customer requirements Monday orders	Final-assembly build schedule (left-hand cell) Monday 1st shift	Monday 2nd shift
14509	A	Black heated	140	500	0
14504	A	Black unheated	110	0	500
14506	A	White heated	120	0	0
14502	A	White unheated	120	0	0
14508	A	Red heated	110	0	0
14505	B	Silver heated	70	0	0
14503	B	Yellow heated	60	0	0
14507	B	Bronze heated	70	0	0
14511	C	Purple unheated	100	0	0
14512	C	Gold unheated	100	0	0
		Total	1,000	500	500

Apogee managers historically had run the cells with large batch sizes because they fundamentally *believed* that the plant ran more efficiently this way. Prior to the lean initiative, a changeover had required a substantial stop of production to change fixtures and ensure that all necessary parts were in the correct positions at the cell to run the next part number. This stop was perceived as a "productivity loss" that could be minimized with large batches that reduced the need for changeovers.

The belief that efficiency was maximized as changeovers were minimized was even more strongly held by managers in the upstream batch processes in paint and molding departments where changeover times were much longer. The paint department always sought to make at least 1,000 items of a given color once it was scheduled and often ran two or three times more of that color once it was running smoothly to avoid changeover problems in the automated cleanroom. Molding then would double the paint department's planned batch size (to 2,000 pieces) to further minimize downtime due to its changeovers, which were even more time-consuming.

At each process step up the value stream, production became less level in relation to customer needs, which was one right and one left mirror for each car moving down the customer's assembly line (and often in a different color or style for the next car). This is shown in the diagram (see *Sample Lot Size Comparisons across the Extended Value Stream*).

Sample Lot Size Comparisons across the Extended Value Stream

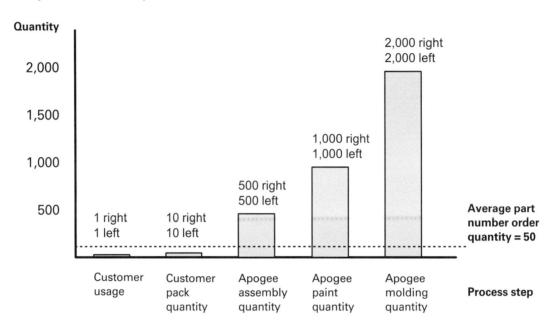

Based on what they had learned, the Apogee team now could see that the previous definition of efficiency had been far too narrow. In order to operate each point along the value stream efficiently—in final assembly, paint, and molding—Apogee was building large batches far removed from customer needs. *Point efficiency* at each process step was producing much larger *system inefficiency* in the form of inventory carrying costs, space requirements, expediting of missing parts, and general management overheads.

The Apogee team now also realized that the biggest gains from reducing batch sizes would be in inventory reductions because *batch sizes determine the replenishment cycle for finished goods*. The team could see this by referring back to their formula (*shown on page 20*) for calculating the minimum amount of necessary finished goods. Shortening the "lead time to replenish" from the pacemaker, which could be done only by reducing batch sizes, would directly lead to a decrease in finished-goods inventory.

For example, if Apogee could build all five A items consistently on a daily basis (rather than once a week) and all five B items every two days (rather than once a week), they could reduce finished-goods inventories for these items by 80% and 60%, respectively (*see Shorter Lead Times Reduce Finished-Goods Inventory*).

Shorter Lead Times Reduce Finished-Goods Inventory

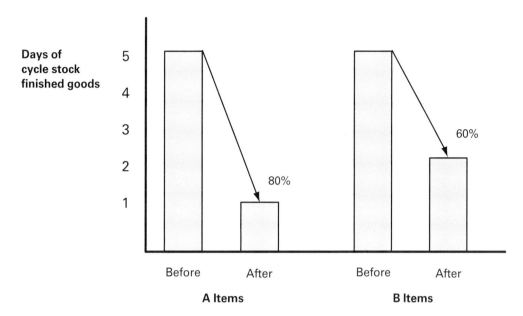

The team now realized that batch sizes at the pacemaker should be minimized rather than maximized, subject to three constraints:

A. Work content differences between products
B. Changeover requirements between part numbers
C. Production pitch interval

A. Work content differences between products

Even though mix and demand may be level at a pacemaker, wide variation in work content for different products moving through a pacemaker assembly cell can create an unreasonable burden for production operators in the cell. Fortunately, Apogee had carefully evaluated work content in setting up its final assembly cells two years earlier to be sure that the work content for all of the products assigned to the cell varied by only a small amount and that no product had work content above takt time.

The easiest item to produce was an unheated, manual mirror requiring 45 seconds of work content per cell station. The most difficult item was a heated, powered mirror requiring 54 seconds of work per station. The weighted average work content for all items moving through the cell was 50 seconds per workstation, which was well under the customer-based takt time of 54 seconds. Work content difference therefore posed no barrier to Apogee in producing in smaller lot sizes in any sequence. And, in fact, reducing the batch size for the heated, powered mirrors with the 54-second takt time from the previous minimum of 500 would make it easier to operate the cell smoothly through the entire shift because it would free up a bit of time to deal with unavoidable variations in output.

As you think about the batch sizes to use in your facility, the move to level pull is a great time to reevaluate the work content of the part numbers going through each operation. Significant variations that have gone unnoticed in a push-scheduling environment, particularly if work content exceeds takt time, will come painfully to light in a tight pull system.

B. Changeover requirements between part numbers

If changeover times were long in the assembly cell (the pacemaker), it also would have been hard to significantly level the pacemaker by reducing batch sizes because a substantial amount of production time would have been lost. Fortunately, two years earlier Apogee also had reduced changeover times at the pacemaker cell to zero. (This included both the time for fixture changes and the time to get materials in place for the next part number.) This meant that there was no minimum constraint on the Apogee batch size due to changeovers. The management had simply failed to translate this improvement into a reduction in batch sizes, choosing to use the freed-up time from shortened changeovers for more production time.

C. Production pitch interval

When work content variations and changeover times do not present obstacles to reducing batch sizes, production pitch determines the maximum extent to which the pacemaker can be leveled by mix. Pitch is a lean concept and is calculated by multiplying takt time by pack quantity (the number of products per container transferred to finished goods from the assembly cell). For example, the completed mirrors produced in the exterior-mirror cell are packed for shipment to the customer in containers with exactly 10 items. This means that with a takt time of 54 seconds, a pack of mirrors ready to ship to the customer is produced every nine minutes, for a pitch (or pitch interval) of nine minutes.

Production Pitch Calculation

Takt time	x	Pack quantity	=	Pitch
54 sec.	x	10 pieces	=	540 sec. (9 min.)

Because pitch is the bridge connecting customer pack quantity with takt time, it makes no sense to produce quantities in the assembly cell that are less than pack size, or in any batch size other than a multiple of pack size. This is because there is no way to convey smaller quantities or partial quantities to the customer. Therefore the minimum batch size in the Apogee final assembly cell was 10 items. (This could change in the future, of course, if the pack size was changed, but this seemed unlikely at Apogee because the pack size was already small. However, in your facility, reducing pack size may be an appropriate step in introducing your level pull system if the current quantity in your packs is very large.)

With clear knowledge of available daily production time, daily customer demand by part number, and minimum batch size, the Apogee team now was in a position to level production by mix. To do this they simply needed to divide the available production time (450 minutes) by pitch (nine minutes) to calculate the number of pitch intervals (50) available to meet demand.

Pitch Interval Calculation

Time available	÷	Pitch	=	Intervals
450 min.	÷	9 min.	=	50 intervals

But how should Apogee allocate the 50 pitch intervals available by part numbers? From their ABC analysis, the Apogee team knew that 60% of demand was for the five A items, 20% for the five B items, and 20% for the 15 C items. They therefore decided to allocate the available production time by category of part number as shown in the chart.

Time Intervals per A, B, and C Items

Total interval	x	% of production mix	=	Intervals per item	(equivalent time)
50 intervals	x	60%	=	30 reserved for As	(9 min. x 30 = 270 min.)
50 intervals	x	20%	=	10 reserved for Bs	(9 min. x 10 = 90 min.)
50 intervals	x	20%	=	10 reserved for Cs	(9 min. x 10 = 90 min.)

Using the pitch-interval logic just described, 60% of the available time was devoted to replenishing A items during each shift. Similarly, 20% of the time (10 pitch intervals) was allocated to B items, and the remaining 20% of the time (10 pitch intervals) was reserved for the C items.

Multiplying pack quantity (10 items) by the number of pitch intervals available for each category of parts shows that Apogee will now produce 300 A items each shift, 100 B items, and 100 C items. This contrasts strikingly with the current situation where the cell produces 500 items of only a *single part number* each shift.

But which A, B, and C items? As a first step, Apogee determined to build each A item every shift because customers ordered each of these items every day. By dividing the five part numbers into the available capacity of 300 for A items, it was easy to see that the A items should be built in batches of 60 (or six pitches) for each part number as a first step.

The next step was to consider batch sizes for the B items. One approach would have been to conduct a changeover between part numbers after every two packs (20 pieces) so the five part numbers could be built every shift. However, Apogee managers were worried about changing over so often in a new system that was untested. And they also noted that B items were on average ordered only every other day. So to get started, they decided to make B items in batches of 50, with the result that every part was made every 1.25 days. (As you approach your own decision on batch sizes, above all be practical. Setting an ambitious target for batch sizes in an immature system actually may lead to slower progress than setting a less ambitious target, successfully achieving it, and then quickly reducing batch sizes again.)

The final step was to set batch sizes for the C items. This was easy. In the block of time reserved for C items on each shift, Apogee would try to produce a whole order requested by a customer and, if necessary, carry this order over into the next available block of time reserved for C items at the end of the next shift. Alternatively, if a customer order was small enough Apogee could produce several different C items during the time interval reserved during a single shift.

With these decisions made, it was easy for Apogee to plan how available production capacity could be distributed by type of item (e.g., A, B, or C) and by part number (e.g., 14509) as shown in the chart on the next page.

(Please note this is not a schedule; the actual schedule will be based on customer withdrawals from finished goods, daily releases, and cell performance.)

Apogee recognized that further leveling of the schedule was indeed possible and desirable so that every part number could be produced every day, then every shift, and then even more frequently if necessary. As a first step, however, leveling production by mix to this extent was both feasible and a tremendous improvement.

Planning Production by Product Type and Part Number

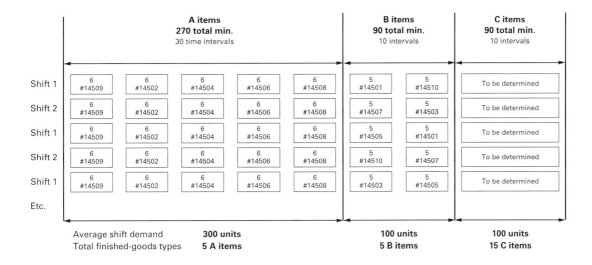

Question 6:
How will you convey demand to the pacemaker to create pull?

Apogee had now answered five of the questions necessary for creating a level pull system. The assembly cell was officially designated as the pacemaker process for the exterior-mirror value stream, and the daily quantity and mix of production were leveled by Production Control.

The next question to answer is how Production Control will convey demand information in the form of production instructions to the assembly cell to trigger pull production. (As we will see, answering this question also will answer the question of how finished goods will be conveyed from the cell to the finished-goods market.)

In lean production, the specific tool for communicating production instructions and for regulating materials conveyance is the kanban. As a downstream process consumes product, signals are sent back to the upstream process via kanban to replenish the amount consumed. When communicating over long distances, some form of electronic kanban often are used in place of simple kanban cards, but within the four walls at Apogee kanban cards are a good way to govern information and material flows.

Rules for Using Kanban

- The downstream process pulls from the upstream process.
- The upstream process replenishes the quantity taken away.
- No defects should be allowed to pass on to the next process.
- Kanban must be attached and conveyed with the part or container.
- No parts can be produced or conveyed without a kanban instruction.
- The quantity indicated on the kanban must equal the actual amount in the container (to ensure accuracy of information).

Note: Readers not fully comfortable with the kanban concept or completely knowledgeable about the different types of kanban should refer at this point to *Appendix: Kanban*.

Types of Kanban

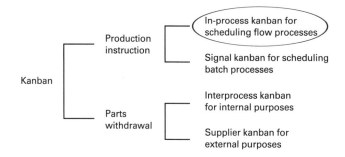

Apogee now needed to create a set of *production-instruction kanban* and specifically *in-process kanban* to communicate production information from Production Control to the assembly cell.

Example of In-Process Instruction Kanban

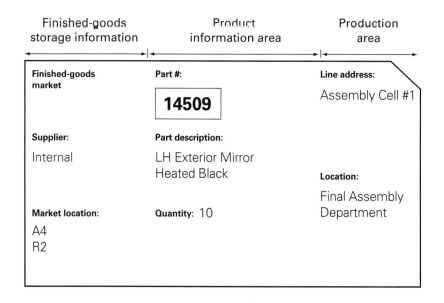

Note that each in-process kanban card is an instruction to produce one pitch (one container) of a given part number, in this case part #14509. Note also that the kanban provides all the information the cell and the material conveyance system will need to make the right amount and send it to the right place:

- Storage location for finished goods in the market (on the left side of the card);
- Production location of the cell (on the right side of the card); and
- Part number, part description, and the quantity to be made (in the center of the card).

Apogee produced these cards for every part number to be run in the exterior-mirror cell. But how should they be sequenced, and at what pace should they be delivered?

One method is to collect all of the cards for the shift and deliver them to the cell at the beginning of the shift, leaving the operators to decide in what sequence to produce products and at what pace to produce. And this is what traditional schedules often do in mass production. However, lean thinkers discovered long ago that a better method, and the one adopted at Apogee, is a heijunka box. This simple device uses time intervals across the top (the column labels) to visually array the production instructions—the in-process kanban—in a way that makes completely clear what part number to produce next and at precisely what time.

Apogee initially set the time intervals across the top of the heijunka box to correspond to pitch intervals of nine minutes. The box was sized to accommodate all 50 pitch intervals in a shift and all of the part numbers to be run during the shift. The box then was loaded by Production Control with all of the kanban cards for items to be produced during the shift. (The box shown on the following page is loaded for the beginning of the first shift on Monday at Apogee.)

Heijunka Box Origins

While we commonly associate the heijunka box with production control for oper-ations, the first application of the heijunka box concept at Toyota actually occurred in maintenance. Many years ago, Toyota managers found it useful to create boxes with hourly time intervals for scheduling preventive maintenance activities. By care-fully timing the work content for each activity, drawing up a work sheet specifying how much time was needed to complete the task, and placing the sheets in a highly visible box with time intervals clearly marked, Toyota helped its supervisors to pace work while remembering to schedule all of the necessary tasks and avoiding scheduling them all at once in a way that might interrupt production. The heijunka device *leveled* maintenance effort (a concept you should try in your own operations).

From this initial application, finer divisions of time evolved and the heijunka idea migrated to pacing production work. The most widespread application has been with suppliers, where the box acts as a tool to pace withdrawals of inventory and to tightly link the output of the supplier's assembly cells (usually manual) to the takt time of Toyota's assembly lines.

Heijunka Box—Nine-Minute Interval

Shift 1	7:00	7:09	7:18	7:27	7:36	7:45	7:54	8:03	8:12
Cell #1	14509	14509	14509	14509	14509	14509	14502	14502	14502

But how are the cards to be conveyed from the heijunka box near Production Control to the assembly cell? Apogee realized that the best method is to deliver the production instructions just-in-time to the cell, ideally one pitch of work (one container) at a time. Doing this introduces a strong sense of pace at the exterior-mirror assembly cell (which lean thinkers often call "takt image.") It prevents working ahead and it also notifies Apogee managers immediately if the cell is not keeping up with the output needed to meet the customer requirement.

The way to introduce just-in-time delivery of instructions is by means of a conveyance operator who picks up the cards at the heijunka box and takes them to the cell. This operator also can push or pull a cart and pick up an empty container needed for holding the finished goods indicated on each kanban being delivered. The container can be taken to the cell attached to the kanban, where the conveyance operator can take away completed products and deliver them to the finished-goods market. Doing this turns out to have the major advantage of precisely linking material and information flow to and from the cell while removing almost all empty containers and finished goods from the area around the cell.

To establish the proper configuration of the conveyance route, Apogee needed to run several time trials linking the in-process kanban, the heijunka box, the empty-container storage area, the assembly cell, and the finished-goods storage area. (You will need to do this as well because there is no secret formula for determining conveyance routes.)

Conveyance Route for Information and Materials

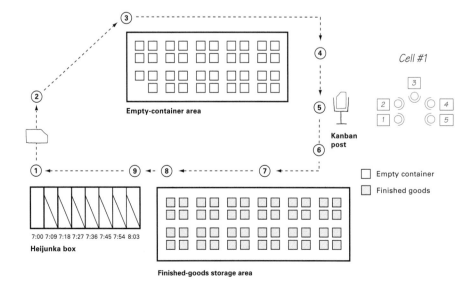

After several time studies, the work elements and flow of the conveyance route were determined to be as follows:

Apogee Conveyance Route Activity List

Step	Activity	Time
1.	Pick up instruction kanban from the heijunka box.	10 sec.
2.	Travel to obtain empty finished-goods container.	1 min.
3.	Obtain empty finished-goods container for delivery to the assembly cell.	30 sec.
4.	Travel time to the assembly cell	30 sec.
5.	Drop off an instruction kanban and an empty finished-goods container at the cell, pick up finished goods from the previous cycle.	1 min.
6.	Travel time to the finished-goods market.	1 min.
7.	Drop off finished goods to their correct storage location.	30 sec.
8.	Log product into finished-goods inventory (e.g., scan bar code).	1 min.
9.	Return to heijunka box for next instruction.	30 sec.
	Total time	6 min. 10 sec.

(Note: The LEI workbook "Making Materials Flow" provides a more detailed description of conveyance route design that you may find useful.)

As established, the route took only six minutes and 10 seconds for the conveyance operator to run—not the nine-minute pitch interval. Also, despite trying to adhere to a takt time of 54 seconds, the assembly cell output often was varying by up to one minute (plus or minus) off the targeted nine-minute pitch time. This fluctuation meant that sometimes the completed pack of finished goods was not ready for pick-up by the conveyance operator. Additionally, one of the stations in the assembly cell had a labeling machine go down for 10 minutes during the trials, an occurrence that rarely happened but in fact did on this occasion.

Apogee's improvement team was learning that even improving the level and pull aspect of scheduling final assembly would not solve all production problems. (Clearly, more point kaizen was needed to stabilize production output from the assembly cell.) More importantly, Apogee learned that they could not react to problems in final assembly as fast as the nine-minute pitch required. The downtime in assembly, although infrequent, meant that a nine-minute pitch interval was faster than they realistically could sustain at their current level of stability.

The team, therefore, made several short-term adjustments. They increased the pitch interval to 18 minutes instead of nine (meaning there would now be two production instruction kanban in each pitch interval per cell in the heijunka box). This way the cell was more likely to have one finished-goods container completed when the conveyance operator arrived, and there was more time for the supervisor to react to minor problems that might occur during assembly. (You are likely to encounter similar issues. As always, be practical as you start your implementation.)

Heijunka Box—18-Minute Intervals

Shift 1	7:00	7:18	7:36	7:54	8:12	8:30	8:48	9:06	9:24
Cell #1	14509 / 14509	14509 / 14509	14509 / 14509	14502 / 14502	14502 / 14502	14502 / 14502	14504 / 14504	14504 / 14504	14504 / 14504

> **Production Instruction Pitch**
>
> How short should your pitch interval be for production instruction? The answer depends upon several factors.
>
> One purpose of the pitch and conveyance withdrawal cycle is to create a sense of pace for a manual assembly line. If you want a strong sense of pull and flow, a short pitch of 10 minutes or less is better. This requires clockwork precision and stability in final assembly, which most companies do not have as they start this process.
>
> The other purpose of the pitch interval is to create a tool to see if production output is maintaining the scheduled amount throughout the shift. The in-process instruction kanban automatically prevents overbuild of inventory because the delivery timing is regulated by the conveyance operator. If the assembly cell is behind, then the pitch interval can become a time bucket for the team leader to manage, determining if the team is keeping up with the schedule. If the cell frequently falls behind, some type of visual signal can be created, such as a red light or accumulation of instruction cards in a collection post at the cell. Toyota calls this type of signal an *andon*, of which many varieties exist. In this way, the interval of pitch can become a visual tool for managers to assess production status and react to problems.
>
> Most companies can't assess production status every 10 minutes at the outset of their implementation. If you calculate a mathematical pitch as short as 10 minutes, you should carefully weigh the benefits vs. the risks. Most companies are probably better off with an initial pitch of 15 minutes to 30 minutes. Any longer than that and you are back to taking hourly or mid-day production counts and not discovering and reacting to problems early enough on the shop floor.

The new pitch left the conveyance operator with only six minutes of work during an 18-minute interval, which obviously was inefficient. As the Apogee team considered this problem, they realized that it now was time to add the second cell for exterior mirrors (for the right-hand versions) to their level pull system. This required a bit of work to calculate the appropriate finished-goods inventory for each part number and to add these part numbers to the finished-goods market. But once this task was done, it was easy to add a section to the heijunka box to regulate the output of both exterior-mirror cells.

Heijunka Box—Exterior Mirror Cells

Shift 1	7:00	7:18	7:36	7:54	8:12	8:30	8:48	9:06	9:24
Cell #1	14509 / 14509	14509 / 14509	14509 / 14509	14502 / 14502	14502 / 14502	14502 / 14502	14504 / 14504	14504 / 14504	14504 / 14504
Cell #2	24509 / 24509	25409 / 25409	25409 / 25409	25402 / 25402	25402 / 25402	25402 / 25402	25404 / 25404	25404 / 25404	25404 / 25404

By combining instructions and conveyance to both cells, the work time for the conveyance operator increased to 10 minutes. This percentage of work still was inefficient, but for the short term the objective was to keep the implementation moving ahead; the team decided to operate the route for two or three weeks while documenting their learning.

As soon as possible Apogee will expand the system to all six cells in final assembly, developing more complex routes that should increase conveyance-operator efficiency. In the meantime, the conveyance person will collect information and document how many times the cells completed their production on time as well as causes of delay. (This information will help prioritize point kaizen in the cell and on the conveyance route.)

Alternative Heijunka Methods

Apogee employed a combination of instruction kanban loaded into the heijunka box with a fixed-time conveyance route. This option represents one way to use the heijunka box, but it is not the only way or necessarily the best for your situation.

For Apogee, the actual ship quantity and mix can change fairly late in the sequence of events (even hours before the delivery truck arrives) and so Apogee must hold finished-goods inventory of A and B items. This scenario dictates a high amount of Production Control observation and intervention to schedule the assembly cell via the heijunka box for each shift. Production Control schedules the heijunka box based upon knowledge of what was removed from finished goods (and now requires replenishment) and any insight into the special C items that might be required that day.

In more stable customer-demand environments and ones where all end items are held in finished-goods inventory, a different method may be better. Instead of putting *instruction* kanban in the heijunka box, use *withdrawal* kanban:

1. From the heijunka box the material handler removes a withdrawal kanban, which is a signal to remove material from a market.
2. The material handler travels to the finished-goods store.
3. From the store the material handler obtains the proper item.
4. While at the finished-goods store, the material handler also detaches a production instruction kanban from the selected item in the store and puts this in a central kanban post in shipping. (This instruction kanban must first be on the item in the market for the system to work. It also must go to the assembly area quickly, either via the material handler or another person to signal a direct replenishment instruction back to the cell.)
5. The material handler delivers the finished-goods item to the dedicated shipping lane for preparation for the upcoming customer shipment.
6. The cycle repeats.

Using withdrawal kanban in this manner has several advantages:
- Delivery of material to the shipping lane is paced and timed evenly across the shift.
- Withdrawal of finished goods automatically triggers replenishment build instructions for the assembly cell.
- The need for frequent direction and intervention by Production Control is greatly lessened because the system self-schedules.

There are, however, several difficulties in this scenario:

- All items in the finished-good store must have an instruction kanban on their container.
- The actual ship schedule must be 100% firm in advance of loading the heijunka box with withdrawal kanban.
- Pacing the delivery of product to the shipping department may not be efficient if only one cell is involved.

Please experiment with both types of basic options for combining kanban with the heijunka box. Variations of each style also exist.

Pacemaker Summary

The Apogee team now had made several important decisions. They had designated a single pacemaker point for the value stream—the final assembly cells. They had identified both the frequency of production based on product mix at assembly (how level) as well as the pitch for delivery of the production instruction. And they had decided to use a heijunka box scheduling tool in conjunction with a fixed-time withdrawal conveyance route to deliver the build instructions to the assembly cells. Instead of adopting the lowest possible pitch of nine minutes for the conveyance route, they temporarily chose a pitch of 18 minutes until the system was more stable.

During the next few days, Apogee built up the instruction kanban needed, set up the heijunka boxes, timed the conveyance cycle for delivery, and trialed the system to debug minor problems. Then they put the system in operation.

Creating the Pacemaker—Keys to Success

- Use the guidelines provided to select the best location for your pacemaker.

- Don't unplug the entire schedule system once you have identified the pacemaker process—there is a lot more work to be done.

- Set a final-assembly lot size after carefully considering work content differences, changeover times, and pitch intervals. As a rule, lot sizes should not vary much from average customer order size.

- To help pace the assembly cells to customer demand, link the pitch interval for instruction to the cycle of conveyance operations.

- Select a pitch size for the pacemaker that is close to your capability to react to problems in operations. If pitch is too short you will not be able to respond within the time period; if it is too long you may not notice small problems until they are big problems.

- Create standardized work for conveyance of information and materials based on the pitch cycle.

Part 4 Controlling Production Upstream

Part 4 Controlling Production Upstream

7. How will you manage information and material flow upstream from the pacemaker?

8. How will you size your markets and trigger withdrawal pull?

9. How will you control batch processes upstream from the market?

Part 4: Controlling Production Upstream

Apogee had made key strides toward their goal of achieving level pull. The Apogee improvement team had determined what items the facility would hold in finished-goods inventory, established how finished goods would be replenished, and identified the single point—the pacemaker—to schedule production.

These decisions coupled with level scheduling of the pacemaker cut overtime and expediting by half. It also improved the ready-to-ship rate from 92% to 97%. But more improvements were both possible and necessary because Apogee had worked only on improving the information flow to the assembly cells. The next challenge was to create production pull *upstream* from the pacemaker process.

Question 7:
How will you manage information and material flow upstream from the pacemaker?

As a first step, the improvement team collected data on production at the pacemaker. For several days the team focused on the cell's ability to deliver finished goods on time to the finished-goods conveyance operator. They discovered that usually the assembly cells kept pace with the production instruction on each pitch interval. But several times per shift the cells failed to produce on time, and the team created a Pareto chart to list the causes.

Assembly Delays

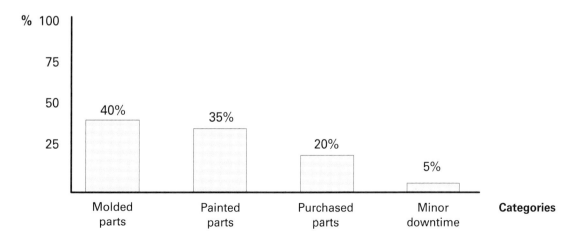

The simple chart showed that the leading cause of delays in assembly was the supply of parts coming from paint, molding, and purchased parts. Downtime in the cells was only a minor contributor to delays.

Why were parts from paint, molding, and purchased parts not getting to the cells on time? Additional investigation quickly identified the fundamental problem: one end of the plant —from the shipping dock back to the assembly cells—now was working with a pull system, while the other end of the plant—from the receiving dock forward to assembly—still was working to a push schedule. Push met pull at the final assembly cells with unacceptable results.

Apogee's Push/Pull Conflict

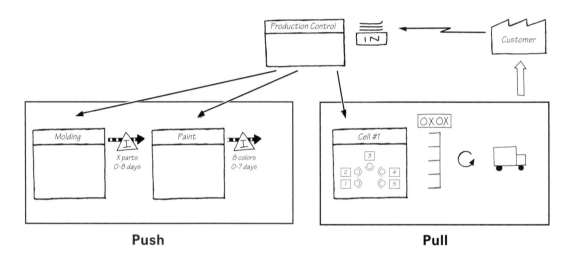

Apogee now needed to regulate production at the upstream processes and to *pull* products forward from paint, molding, and purchased-parts storage.

In an ideal world, upstream production of parts would be conducted one piece at a time at the same rate as final assembly, and parts would flow directly to the pacemaker without interruption. But at Apogee this clearly was impossible given the batch nature of the paint and molding processes and the long distances to the external suppliers of parts.

Because the upstream processes had different operating patterns compared to final assembly cells, significant changeover times, and in the case of purchased parts significant travel distances, a formal market mechanism was needed to properly regulate flow.

Exterior-Mirror Process Review

Process	Supplies exterior-mirror value stream?	Batch process with changeovers?	Long distance to next area?	Different shift pattern?
Paint	Yes	Yes	No	Yes
Molding	Yes	Yes	No	No
Purchased parts	Yes	No	Yes	No

The Apogee team realized that they would need to create *controlled markets* to buffer the flow of the three main types of parts upstream from the assembly cells because careful management of this inventory would be the only way to guarantee smooth output in the pacemaker cells. Additionally, a tool known as the *withdrawal kanban* would be needed to regulate the movement of parts between these markets and the assembly cells.

Based on this decision, Apogee's improvement team proposed to incorporate into the exterior-mirrors value stream *markets* for molded, painted, and purchased parts, with pull loops between each downstream process and the upstream parts markets.

Apogee's concept in implementation was to store a minimum amount of inventory for each A and B item in the assembly cells and to store the remainder (including the parts for C items) in the markets.

Making Materials Flow

For detailed information on establishing markets, especially purchased-part markets—including the Plan for Every Part (PFEP) and examples of material handling—please refer to the LEI workbook, *Making Materials Flow*. The focus of this workbook is on the system-level mechanics required to implement the level and pull methods that make material-handing tasks effective.

Apogee took the following steps to set up the markets, which they located together to aid management and minimize the length of conveyance routes *(see plant layout on page 52)*.

- A spreadsheet was created listing every part used to assemble exterior mirrors. This spreadsheet—referred to as a *Plan For Every Part* (PFEP)—listed every relevant detail about each part and its path through Apogee. This helped Apogee confirm that the right parts were in the market, as well as estimate size requirements for racks and shelves.

- Space was marked off on the production floor for the market in one central area, and racks and shelving were constructed to hold the material. A precise address system was created (including the address of the delivery locations at the assembly cells) so that a given part was stored in only one, precisely defined location.

- Over the next weekend, Apogee's improvement team moved the material from many locations into the central market location and tested delivery to the pacemaker cells using a conveyance route with a tugger pulling carts of material.

Markets and Point-of-Use Addressing

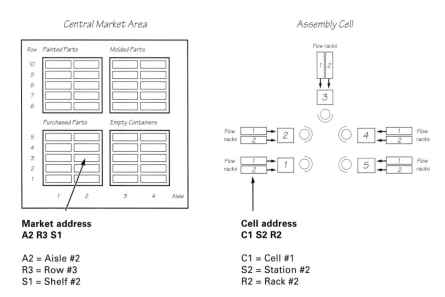

Market address
A2 R3 S1

A2 = Aisle #2
R3 = Row #3
S1 = Shelf #2

Cell address
C1 S2 R2

C1 = Cell #1
S2 = Station #2
R2 = Rack #2

As noted, Apogee established a basic addressing system for the assembly cells and the central markets. This information will be needed for the withdrawal kanban and helps material handlers identify where exactly to go for delivery and pick up of material.

Locating the Markets

Different options for locating markets should be considered. Traditional Toyota practice has been to locate them directly at the end of the producing process. This helps the process see exactly what and how much is being made of each item. Downstream customers (i.e., processes) are expected to come and pull what they need on a regular basis (either on a fixed-time or fixed-quantity conveyance route). The producing process then simply replenishes whatever is taken away. This method may work well for you as a corrective action if you detect a tendency to overproduce and push items downstream, provided of course that you have room to store all parts near the process.

Facilities that have many internally produced parts, which is the case at Apogee, or that have layouts that can't accommodate all items at the end of the producing process can try centrally located markets. These ideally locate all components in proximity to both the producing processes and the downstream operation (the final assembly cells in Apogee's case). This option may increase overall material-handling efficiency and minimize the distance traveled on the parts withdrawal loop.

Alternatively, in plants that do only assembly and bring many parts in from suppliers on large pallets, a good location is near the receiving dock. This market position will favor ease of unloading and storage of material into a market and minimize the conveyance distance for the receiving team, although it may lengthen the travel distance to the assembly cells.

Whichever option you chose, be sure the following rules apply:

- Make clear that the producing process has ownership of the inventory it produces.
- Ensure that downstream processes pull what they need when they need it on a regular basis.
- Ensure there are signals exchanged between the processes regarding what has been taken away (a topic we will cover shortly).

Updated Plant Layout with Market Area

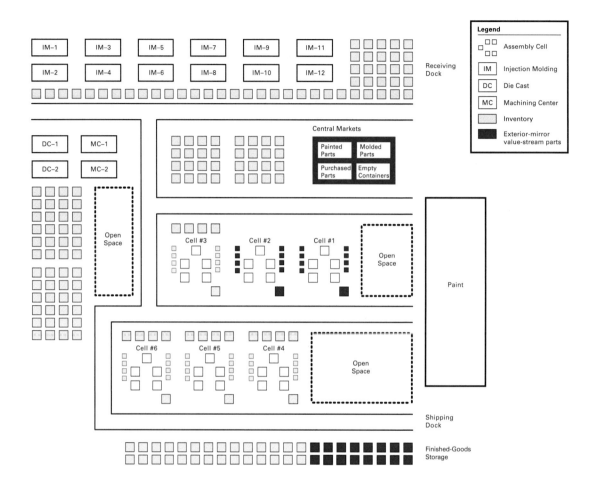

Question 8:
How will you size your markets and trigger withdrawal pull?

Apogee now needed to implement methods to regulate and pace the flow of information to the central market and the flow of material from the market to the assembly cells. This will ensure the right parts are available at the right time. Specifically, Apogee needed to employ *withdrawal kanban*. These govern *what to deliver* to the assembly area from markets and support the assembly of product as dictated by the production instruction kanban arriving at the cells from Production Control.

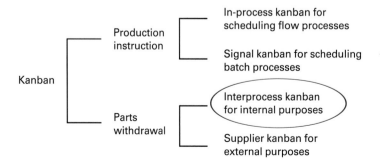

Apogee employed *interprocess kanban* between the assembly cells and the parts market. (The other type of withdrawal kanban is the *supplier kanban*, which controls the information loop from the purchased-parts market back to external suppliers. Apogee will need to deal with the supplier-kanban loop eventually, but we will restrict the current exercise to replenishment of parts *within* Apogee's facility, as regulated by the interprocess kanban.)

Example of Interprocess Withdrawal Kanban

Supplier information area	Product information area	Point-of-use area
Supplier code: ABC	**Part #:** X2174	**Line address:** Station #1, Flow Rack #2
Supplier: Ajax Springs	**Part description:** Spring	
		Line location: C1, S2, R2
Market location: A2, R7, S1	**Quantity:** 40	

Apogee needed to complete three tasks to successfully trigger pull from the assembly cells to the central market:

A. **Set a standard amount of inventory of each part to hold at the assembly cells based on the nature and frequency of the conveyance route.**

B. **Create a separate withdrawal kanban for each container stored at the cells.**

C. **Determine the right amount of inventory to hold in the central market.**

A. **Set a standard amount of inventory of each part to hold at the assembly cells based on the nature and frequency of the conveyance route.**

How much inventory to hold at the assembly cells (called "inventory line side") depends on how frequently this inventory will be replenished, the minimum pack size for parts, and the operation of the delivery route. For example, if material is delivered only once per shift, then a minimum of one shift of inventory needs to be kept at the assembly cells. But with eight or more hours' worth of material stored line side, the cells will look more like a storage facility than a production site. In addition, finding the right item will be difficult, floor space will be wasted, and the large area required for each cell will lengthen the conveyance route.

Instead, material could be delivered in small containers and at a much quicker rate. The majority of the 30 parts that Apogee stored line side per cell came in containers holding six to 25 items (exceptions were fasteners, which have many more items per container). Apogee theoretically could continually deliver all basic items to the cells as the parts in a single container of a given part number were used. One container always would be at the cell, and another would be in circulation with the parts withdrawal conveyance operator.

Lean facilities with only a small number of parts and tight assembly areas can adopt this *two-bin* style of conveyance. However, with many part numbers and small quantities inside each container, two-bin conveyance can lead to excessive material handling due to the need for ultra-frequent runs by the conveyance operator.

Because Apogee has many parts line side and the internally molded items have as few as six items per container, a conveyance method was needed that was somewhere between the two-bin style and the traditional method that placed one shift of inventory at line side.

To achieve this, the Apogee team first needed to determine the nature of the conveyance route and conducted a number of small conveyance time trials between the cells and the central market, using mock withdrawal kanban on index cards.

Withdrawal Material-Handling Loop

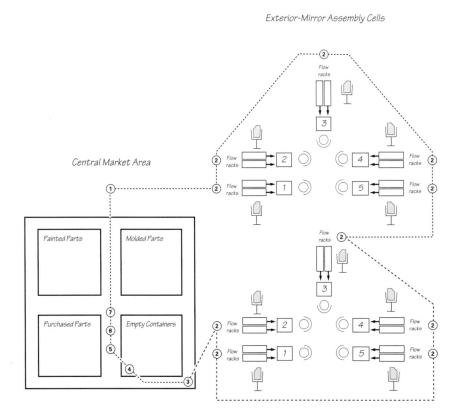

54   Creating Level Pull

For experimentation and learning purposes, Apogee first placed four containers of every item at line-side for a total of about 120 containers for each exterior-mirror cell. Apogee knew this was too much inventory for some part numbers and too little for others, but chose this level for their trial.

Then Apogee placed the mock withdrawal kanban on each of the 120 containers stored line side at each cell with the basic information necessary to identify the part number (*see Example of Interprocess Withdrawal Kanban on page 53*).

Simple kanban posts were erected at each station of the assembly cell and production operators were told to detach the withdrawal kanban each time they began consuming a container of material such as nuts, bolts, wires, brackets, and housings to build mirrors. The detached kanban then was placed in the kanban post.

After several trials between one of the assembly cells and the central markets, Apogee determined the following time and work standards for the material-handling route:

Standard Work Flow and Times for Withdrawal Material-Handling Loop

Step	Standard work elements	Conveyance time estimate
	Cell operator withdraws kanban card from a container upon removing the first part out of the container. The withdrawal kanban cards collect inside a kanban post at each cell station.	N/A
1.	The material handler travels from the central market and arrives at assembly cells.	2 min.
2.	The material handler visits and picks up all the withdrawal kanban at each station of each assembly cell as well as any empty containers, delivers parts to cells, and returns to the central market.	4 min.
3.	Once back at the store, the material handler sorts the cards for the best pick order.	15 sec.
4.	Empty containers picked up on the route are dropped off in the specified location in the market.	1 min.
5.	Any necessary bins for small items, such as nuts, bolts, screws, washers, etc., are obtained.	1 min.
6.	Parts are picked from the store and placed on the material-handling cart (an average of 10 withdrawal items each at 20 sec. per pick).	3 min. 20 sec.
7.	The withdrawal kanban cards are placed into their respective containers.	15 sec.
	Total route time	11 min. 50 sec.

Apogee calculated that on average 40 withdrawal kanban cards would be triggered per hour of production by the two assembly cells for exterior mirrors. The actual travel time between the market and cells was only four minutes. This meant that each conveyance and delivery cycle would contain the four minutes of travel time plus a variable amount of work represented by the number of the kanban cards per trip.

But how frequently should the withdrawal material-handling loop be operated? The Apogee team decided to start with a 15-minute interval to gain experience with the new system.

They calculated that about 10 containers would be required by the two cells every 15 minutes (40 withdrawals per hour divided by four loop operations per hour) and calculated that this would require seven minutes and 50 seconds of work in addition to the four minutes of travel time between the parts market and the cells. This meant a total of 11 minutes and 50 seconds of work on each loop. Trial runs quickly showed that the conveyance operator using a tugger to deliver parts to both cells could easily manage this workload. (Eventually, an additional cell can be added to this loop to use the three minutes and 10 seconds of extra time.)

Every time a set of withdrawal kanban was picked up from the cell (approximately 10 kanban), the material was replenished on the next delivery cycle. With a 15-minute route, Apogee could store as little as 30 minutes of inventory at the assembly cells—far less than they had in the past. But because Apogee did not want to risk shutting down the assembly cells while the withdrawal pull system was being implemented and refined, they conservatively chose to hold one hour of material at the assembly cells and reduce this amount over time.

Next Apogee determined what one hour of material equated to for every part number at the cell. For some items it was one container (e.g., fasteners) and for others it was several containers. Flow rack lengths and positions at the cells then were adjusted and finalized based upon this decision.

Two Types of Conveyance Routes

So far we have examined one type of conveyance route—*fixed-time variable-quantity*. With this method the conveyance operator performs a precise sequence of steps standardized by time, much like standardized work for an assembly cell operator. The time for the route is fixed, but the amount of material moved during a given withdrawal interval will vary based on the previous usage by the production area.

In contrast to fixed-time variable-quantity conveyance, *fixed-quantity variable-time* conveyance uses a quantity-based trigger to signal the need for material movement between locations. The time to deliver the quantities fluctuates as needed, as does the route traveled. Since periodic route times cannot be created with this method, the key to success is a clear visual and/or audio signal that identifies when material needs to be moved to a production process.

Lean operations often need both styles of conveyance. One size does not fit all. When the goal is to frequently move material to a production area in conjunction with a pitch interval (as at Apogee), fixed-time conveyance is best. When intervals are highly infrequent or parts are exceptionally heavy or cumbersome, such as automotive windshields, fixed-quantity conveyance may be better.

B. Create a separate withdrawal kanban for each container stored at the cells.

The Apogee team now was ready to create formal interprocess withdrawal kanban for each container at the assembly cells in the exterior-mirror value stream. The system can work only if there is a physical withdrawal kanban for each item stored line-side at the assembly cells (*see Example of Parts Interprocess Withdrawal Kanban on page 53*).

Each kanban must contain address information on the storage point for the part in the central market and address information for where material will be delivered in the production area. Apogee had already completed the former task, and now transferred both types of information to withdrawal kanban.

Apogee's exterior-mirror parts withdrawal loop consisted of 30 different part numbers with one to six containers stored line side for each, representing one hour of inventory per item. The total number of withdrawal kanban in the loop for the exterior mirror cells was 110 cards. (Part 5 will show how Apogee can reduce the number of cards and containers in circulation as improvements occur.) For control and management purposes, Apogee incorporated this information into its parts-tracking spreadsheet. (Often companies use their PFEP for this purpose.)

C. Determine the right amount of inventory to hold in the central market.

The final issue for the Apogee team to resolve for effective management of the central market was the sizing of inventory at the markets. As parts were moved from many locations in the plant into the central markets, the Apogee team discovered that the plant had anywhere from one day to two months' consumption of different part numbers! They obviously needed a more precise rule for setting inventory levels in the markets.

The basic rule they developed was similar to the formula used for finished goods on pages 20 and 21. The demand variation, though, was much lower than with finished-goods inventories because external demand variation from the customer already was being absorbed by the buffer stock in the finished-goods market. (This is how adding to inventories at one point in a facility—in this case in finished goods—can reduce total inventories in the facility.)

Apogee also had leveled the assembly cells, which reduced internal variation during a normal replenishment cycle. Because of this, a lower buffer-stock factor could be assumed since this variation also would be absorbed by the safety inventory in finished goods.

Market Inventory Calculation

	Average daily demand x Lead time to replenish (days)*	**Cycle stock**
+	Demand variation as % of Cycle stock	**Buffer stock**
+	Safety factor as % of (Cycle stock + Buffer stock)	**Safety stock**
=		**Market inventory**

Market Inventory Calculation for Part #14117 (Painted Bracket)

	100 x 5*	**Cycle stock**	500
+	10%** of 500	**Buffer stock**	50
+	10%*** of (500 + 50)	**Safety stock**	55
=		**Market inventory**	605

   * Lead time for paint process to replenish this item
  ** Two standard deviations of internal variation for this part number
 *** Average scrap rate and downtime for this part number

Apogee used this calculation and logic for all items in the central market. At the end of the first full week of running the exterior-mirror cells with the parts withdrawal kanban route, Apogee's ready-to-assemble performance for this value stream improved from 75% to 91%.

Apogee Ready-to-Assemble performance

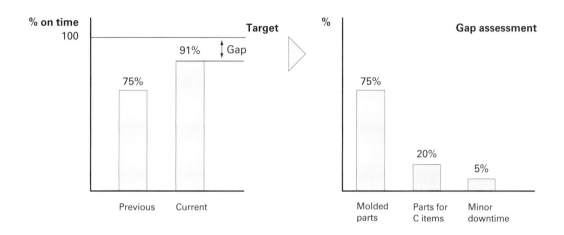

Within the first week of trial operations, the withdrawal kanban loop largely eliminated part shortages at the exterior-mirror assembly cells, which ran overtime on only one day and for only 30 minutes. Apogee now was shipping and delivering to the customer with 100% on-time performance. The remaining problems in the exterior-mirrors value streams were related to getting a few internal parts to the central markets exactly when they were needed and in the right amount, as indicated by withdrawal kanban. Although still short of their final goal, the Apogee improvement team took a moment to celebrate their success!

A Pareto analysis (*as shown above*), based on data collected during the week by the parts-withdrawal material handler, provided insight on the remaining problems. The most important insight was that molded parts still were not always in the market when needed by the parts-withdrawal material-handling loop and, thus, were not ready for assembly. The assembly cell now was pulling from the market based upon a fixed time withdrawal cycle, but the flow of material into the market from the fabrication processes still was regulated by a push schedule. The conflict between push and pull had migrated once more and now was focused on the batch processes.

Managing the Flow for C Items

When creating the central parts market, the Apogee team faced an important choice with regard to parts for C items. The team already had decided that no inventories of C items would be maintained in the finished-goods store. Now the team needed to decide whether to build C items to order from the molding process forward or hold all of the part numbers needed for C items in the central market and assemble the items when an order came in.

Apogee realized the latter option would be much easier to manage—particularly at this stage in the implementation of a level pull system—than manufacturing all of the needed parts to order each time. Apogee therefore purchased a month's worth of the purchased parts for C items and molded and painted a month's usage of the internal parts. This was a much higher level of parts inventory, measured in days of usage, than for the A or B items, but the actual number of pieces (and dollars of inventory) was not that large (e.g., generally fewer than 100 to 200 pieces per part number).

For your situation you should consider all possible solutions regarding production of C items. If there is enough volume and associated business margins, the C items may belong in their own value stream or dedicated assembly cell.

If, however, C items must be made using the same set of production assets as A and B items, you need to decide between two options: Either make each C item completely from the first process in the value stream on through to finished goods *or* make each C item from some midpoint in the value stream where you hold partially completed C items. This latter option is what Apogee chose to follow, and, as long as it does not result in a large amount of slow-moving inventory, is an acceptable solution.

Question 9:
How will you control batch processes upstream from the market?

The last question Apogee needed to answer to complete their new production control system was scheduling of the processes producing batches of parts upstream from the central markets. As shown previously in the Pareto chart, the molding process now was the largest source of problems in terms of supplying material on time to the assembly cells for exterior mirrors. Therefore, the team decided to start with these parts.

Apogee Pull Against Push

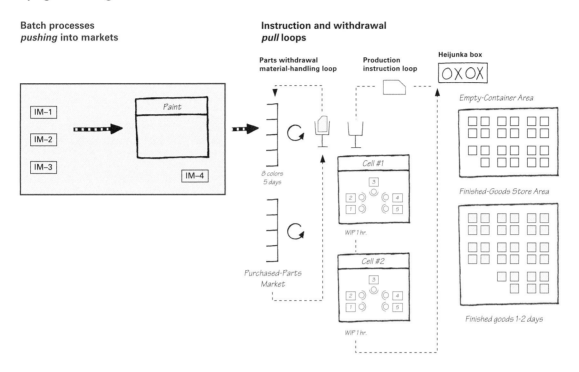

For scheduling batch processes lean thinkers employ a set of tools known as *signal kanban*. The common feature of these tools is that they aggregate the small increments of demand communicated by withdrawal kanban—which arrive at the central market from continuous processes downstream (final assembly in Apogee's case)—into larger batches of demand corresponding to feasible run lengths for the batch processes.

Types of Kanban

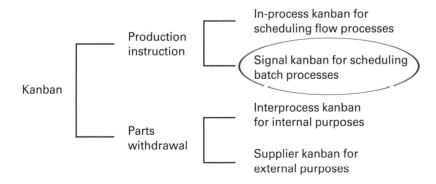

For Apogee's molded parts, the team decided to use a type of signal kanban known as *triangle kanban* (two other options are described at the end of this Part).

Apogee batch sizes in the molding department were a *minimum* of 2,000 units (and often far higher). With average cycle time across all the machines of 40 seconds and only one part produced on each cycle, a "small" run of a part number for Apogee still required more than 22 hours or close to three shifts.

Four machines in Apogee's molding department were dedicated to the value stream for exterior mirrors. Three of the machines made the visible exterior components, including the base and the mount, which were painted as well. The fourth machine produced three molded but unpainted items for both right- and left-hand mirrors that were inside the mirror housings and invisible to the customer. For simplicity, Apogee decided to start signal kanban implementation with this machine and its three part numbers.

Apogee's Production Control Department previously scheduled molding by using an economic order-quantity (EOQ) model to determine the efficient lot size to run in production. As the team investigated the actual run time and changeover patterns, however, they found that supervisors ran most part numbers longer than called for by the EOQ-based schedule. They were measured on machine utilization, and the best way to increase utilization was to do as few changeovers as possible.

To change the molding-machine operating paradigm the Apogee team eliminated the traditional efficiency measurements for the trial period and told operators simply to make parts in the amount and sequence called for by the new triangle kanban. This subtle but important change significantly reduced the incentive for the supervisors to encourage overbuilding of molded product.

Production Control Using Triangle Kanban

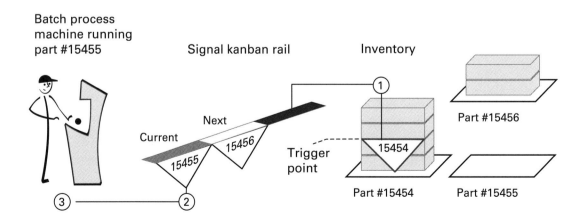

With the triangle (signal) kanban, the removal of inventory from the market triggers replenishment through the following process:

1. Inventory is picked from the storage location and taken to assembly.

2. Once a predefined trigger level is reached in the central market (after a number of withdrawal kanban and containers of material have been picked) the signal kanban for the part number is removed from the market and placed on a rail in front of the molding machine. (Note that there is only one triangle kanban in the system per part number.)

3. The operator follows the instructions on the signal kanban and builds precisely the required *lot size* indicated on the triangle kanban, which is the amount needed to replenish inventory in the market.

To implement this type of system, Apogee had to determine the maximum amount of inventory to have in the market (the lot size) and the level of inventory that triggers the release of the triangle kanban for conveyance to molding. An example of a triangle kanban is shown below.

Example of Triangle Kanban

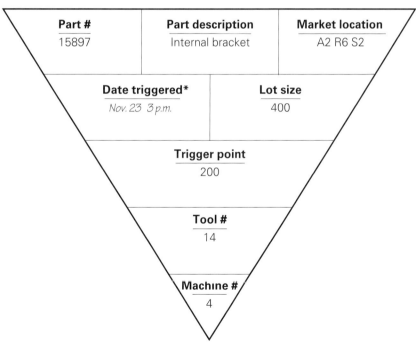

* The point in time at which the kanban is placed on the signal kanban rail. This aids production management in assessing status of the process and reduces the possibility of getting kanban out of sequence.

The improvement team started the conversion process by gathering basic information on customer demand and the #4 injection-molding machine.

Injection Molding Example—Machine #4 Data

Part #	Average demand* per day (pieces)	Cycle time per piece	Average scrap rate	Required run time per day**	Average changeover time
15897	400	20 sec.	1.5%	136 min.	55 min.
15898	600	22 sec.	1.3%	223 min.	55 min.
15899	1,000	20 sec.	1.5%	339 min.	55 min.
Total	2,000			698 min.	

* To make 1,000 left-hand mirrors and 1,000 right-hand mirrors in assembly cells, as dictated by customer demand
** Adjusted for average scrap rate and rounded up to nearest whole number

The team then took the following four steps to implement the triangle kanban pull signals.

Step 1: Determine time available for changeover work.

Step 2: Set the number of changeovers per day.

Step 3: Determine production lot size.

Step 4: Specify a trigger point for reorder.

Step 1: Determine time available for changeover work.

Time Available for Changeover Work

Total 1-shift production time available (net breaks and lunch)		450 min.
Number of shifts	x	2
Time available for production on 1 machine during 1 day	=	900 min.
Time required per day to meet average demand*	-	698 min.
Net time available for set up and changeovers per day	=	202 min.

* From Machine #4 Data

On an average day, with a volume of 2,000 total pieces needed from machine #4, there are 202 minutes available for nonproduction activities, including set-up and changeover work.

Creating Level Pull

Step 2: Set the number of changeovers per day.

With an estimate of the time available for nonproduction work, the team could establish a target for the number of changeovers per day.

Possible Changeovers per Day (Two Shifts)

Nonproduction time available		202 min.
Average downtime per day (not including set-up and changeover times; each shift had 15 min. downtime)	−	30 min.
Time available for changeover work on 1 machine during 1 day	=	172 min.
Average changeover time	÷	55 min.
Possible number of changeovers per day	=	3*

* Rounded down to nearest whole number from 3.13

The team subtracted the average downtime per day to determine how much time was available for changeovers, and then reversed Apogee's long-standing management practice: Instead of seeking to minimize changeovers, the Apogee team determined the *maximum* number of changeovers permitted by the time available and set the number of changeovers per day at this number (three changeovers).

Step 3: Determine production lot size.

Given that Apogee would now execute three changeovers per day, the team needed to set the correct lot sizes (the amount of parts called for by the triangle kanban) and the timing of changeovers.

There are two basic ways in lean manufacturing to set lot sizes with triangle kanban:

One is known as the *fixed-time variable-quantity* method. It calls for producing a constant (fixed) interval of demand for each part number run. For example, one shift's worth of customer demand might be run for each part number. Because demand per unit of time (e.g., per shift) differs by part number, this means that the process will make a different (variable) quantity of parts for each part number run. (In thinking about this, be careful not to confuse machine run time with the demand interval.)

The other option calls for running the same (fixed) quantity of parts for each part number run, producing parts for varying intervals of demand for each part number. For example, a process might run every part number for three hours before changing over to the next part number. This would create lots of perhaps a day's worth of consumption for some parts and two days' for others. (Note that with this method production is adjusted to demand by running some part numbers more frequently than others.)

The first option builds lot sizes more closely linked to customer demand and leads to lower inventories. The second option may be easier to manage because the pattern of changeovers is steady and predictable. Because Apogee felt its molding management was equal to the challenge and because it wished to supply the central market for molded parts in a highly responsive manner, it chose to implement the fixed-time variable-quantity method for the triangle kanban lot-size calculation. A lot size of one shift's worth of demand for each part number was selected, as shown in the diagram.

Fixed-Time Lot Sizes and Changeovers

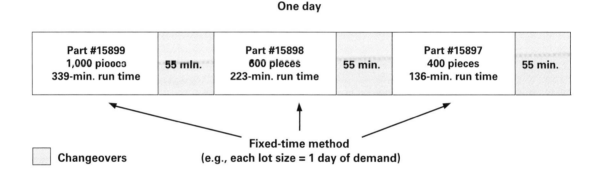

For molding machine #4, with three products and three changeovers each day, the production lot size can simply be set at one day's average demand (three products ÷ three changeovers = one day of demand).

Note that with more frequent changeovers the lot size called for on the triangle kanban naturally becomes smaller. For example, if six changeovers per day were possible, as changeover times are reduced then only one-half day of inventory would be needed per part number.

Lot Sizes with Fixed-Time Variable-Quantity Method

Part #	Lot size	x	Demand per day (pieces)	=	Lot size (pieces)
15897	1 day	x	400	=	400
15898	1 day	x	600	=	600
15899	1 day	x	1,000	=	1,000

Because each part is stored in containers of 10 parts and each part number is being produced to meet consumer demand every day, Apogee will need space for 200 containers in the market to hold the 2,000 pieces of total inventory for the three required part numbers. Previously, when Apogee was producing large lots with infrequent changeovers to obtain high machine utilization, there often were 5,000 to 10,000 pieces per part number (5 to 10 days) of these items at various storage locations in the facility, yet these still were not consistently delivered on time to assembly!

With lot sizes determined, the operator of the molding machine will know precisely how many parts to make each time a signal kanban for a lot of parts arrives at molding. This leaves as the only remaining question, "How do you set the trigger point in the central store for sending the signal kanban to the molding machine?"

Step 4: Specify a trigger point for reorder.

Setting of the trigger level for inventory replenishment with a signal kanban is simple, but it does require several steps: You need to determine the elapsed time necessary for the molding machine to make a standard batch of the product (including changeover time), add the time needed to deliver the first container of parts to the market, and then divide by takt time. (Adding a buffer for external-demand variation is not necessary because this is now covered by the buffer stock in finished-goods inventory and the internal schedule is level.)

For machine #4, the time to replenish once the triangle is triggered can be estimated as follows:

Lead Time to Replenish

Part #	Run time*	+	Changeover time	+	1st container time**	=	Total lead time to replenish
15897	136 min.	+	55 min.	+	10 min.	=	201 min.
15898	223 min.	+	55 min.	+	10 min.	=	288 min.
15899	339 min.	+	55 min.	+	10 min.	=	404 min.

* Run time equals the part cycle time multiplied by the lot size.
** Assume 10 min. to make enough parts to fill one container and to convey it to the market for use.

Of course, each of these individual replenishment times assumes that no other product is running in the machine or waiting to run, and this is clearly not realistic. With three part numbers running on machine #4, it is normal that one part number will be running, one part number will be waiting on the signal rail to run, and one part number will be in the central market not yet triggered. It is statistically unlikely that all three part numbers ever will be triggered at once because there is always some lag between them. Therefore, Apogee assumed that only one triangle kanban will be in front of another when it is triggered. For an extra margin of safety, Apogee assumed that the triangle with the highest demand for molded items (and the longest run time) is at the front of the triangle kanban rail and set to run.

The longest replenishment time for machine #4 therefore is 404 minutes, since that is the longest lead time possible before replenishment of the next part number can begin. Of course, this extreme assumption is appropriate only for two of the three part numbers—the calculation for the third part number can use the second-longest lead time, since a part number cannot be in front of itself.

Part Takt Time Calculation

Part #	Time available per day	x	Seconds	=		÷	Demand per day	=	Part takt time
15897	900 min.	x	60	=	54,000 sec.	÷	400	=	135 sec. (2.25 min.)
15898	900 min.	x	60	=	54,000 sec.	÷	600	=	90 sec. (1.5 min.)
15899	900 min.	x	60	=	54,000 sec.	÷	1,000	=	54 sec. (0.9 min.)

Trigger Point Calculation

Part #	Longest lead time*	÷	Part takt time	=	Trigger point** (number of pieces left in market)
15897	404 min.	÷	2.25 min.	=	180 pieces
15898	404 min.	÷	1.5 min.	=	270 pieces
15899	288 min.	÷	0.9 min.	=	320 pieces

* The longest possible lead time for another product that theoretically could have entered the run queue for that process in front of the required product
** Rounded up to nearest whole number of 10

For each of the three part numbers running through molding machine #4, Apogee was able to calculate a new lot size as well as the signal point at which they should start replenishment for each item. For part #15897 the market would start with 400 pieces of inventory for this item (40 containers). The triangle kanban trigger point would be placed at 180 pieces or 18 containers of product. When the inventory decreased to this level, the kanban would be moved to the injection-molding machine to act as a signal for replenishment. The replenishment calculation shows that molding can produce the product and return it to the market just before the remainder of the inventory in the market is consumed.

Summary of lot size and signal points for signal kanban

Using these formulas and rules, Apogee established the new required inventory quantities for machine #4 and then for each of the other molding machines in the value stream that supplies parts to the central market for exterior mirrors.

For each part number in the value stream, Apogee created one signal kanban and also reorganized the material in the market. As inventory was taken away from the central market by the withdrawal material-handling loop, it triggered replenishment build instructions for the molding machines.

The improvement results were drastic: from 50% to 90% of the previously existing material in the market could be eliminated. More importantly, the replenishment time in every department along the value stream now could be measured in hours instead of days as before. Apogee realized that this was a major breakthrough in terms of enhancing the production system's capability to deal with changes in demand and the ability to supply products on time in the future.

Looking ahead, Apogee will be able to further reduce replenishment time in molding if changeover times can be reduced from the current average of one hour. Indeed, if set-up times can be taken down to single minutes, as many molders are now doing, it should be possible to add a fourth changeover on machine #4 and cut lot sizes by another third.

Notes on Trigger Points

Batch process running many part numbers: If Apogee had needed to run more part numbers through machine #4, then the lead time to replenish any given part would need to be set higher to compensate for the other parts numbers that may be in front and waiting to be run. In general, no more than half the kanban ever will be in front when the production cycle is running in normal conditions. A conservative estimate for the number of kanban on the rail would be to divide the number of signal kanban for that asset by two and then subtract one kanban: (total kanban ÷ 2) – 1 = kanban on the rail.

Unstable production: In a less stable production environment, you may want to establish an earlier trigger point. The Apogee explanation is a lower limit calculation that could cause the market to run down to the final container of product in some rare instances when the full lead time to replenish the previous item is required.

Other Choices in Signal Kanban for Batch Processes

Triangle kanban is the standard method used in lean manufacturing to schedule a batch production process. However, it is only one type of signal kanban. The other common types are *pattern production* and *lot making*. Usage depends upon your individual circumstances, capability, and the type of process you are operating.

Pattern Production: Pattern production, as the name implies, creates a fixed sequence or pattern of production that is continually repeated. For example, in an eight-hour cycle part numbers are always run A through F. The difficulty of your changeovers (for example from light to dark paint or from light- to heavy-gauge steel) may dictate this order for your processes. Inventory in the central market is a function of the length of the pattern-replenishment cycle. A one-day pattern implies one day of inventory must be kept in the market. A one-week pattern requires one week of inventory.

There are several advantages to pattern production. It is simple in that every part is made every cycle (which you must determine based upon average demand), and the changeover sequence is standard and predictable. Companies with high downtime and other problems may find this a much easier way to begin than with the triangle kanban. However, as with any tool, the pattern and quantity run must be analyzed and updated over time to best match production.

The main disadvantage of pattern production is that the sequence is fixed. You can't jump from making part C to part F. Instead, you must finish the sequence. Also there is little incentive to improve changeover time and reduce inventory. The pattern just keeps repeating to replenish material in the market location.

Apogee utilized a form of pattern production to schedule its paint department. A color sequence on a rotating wheel was used to indicate what color to paint next. The basic paint sequence was set for the whole week, for the 50 colors required each week. One week's worth of material for each part number was kept in the central market. The wheel also contained several planned open spaces that were reserved for items such as infrequent service part items and conducting planned maintenance tasks.

As the sequenced color came up to run, the paint supervisor confirmed the amount of inventory in the central market that had been consumed during the previous week and scheduled this amount. Although a visual system, the method was highly effective. If you have the capability, you can use instruction kanban and attach them to each container of painted items in the central market. As the items are consumed during the week, the instruction kanban are detached, collected, and taken back to the paint shop. When the sequence indicates it is time for the color to be run, the cards are counted to determine the paint quantity for that cycle.

Lot Making: Lot making is another effective method, although a more difficult way of scheduling batch processes. Unless you are already adept at the two previous forms of signal kanban or have no other alternative, I suggest you avoid it.

With lot-making methods, a physical kanban is created for every container of parts that is controlled in the system. This can equate to hundreds of cards depending upon on how many part numbers and containers are in the central market. As material is consumed from the market, the kanban are periodically detached and brought back to the producing process and displayed on a *batch board* that highlights all part numbers and displays an outlined shadow space for each of the kanban cards in the system.

Lot-Making Batch Board

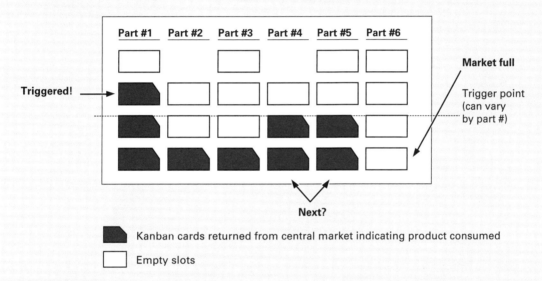

A returned kanban card placed on the board in the shadow space indicates inventory has been consumed in the market; unreturned cards represent inventory still in the market. As predefined trigger points are reached, the production operator knows to begin making product to replenish the material in the market.

The advantage of the lot-making method is that it allows information to come back to the production process more often. It signals what actually has been taken away and uses smaller increments than with the signal kanban. It also provides a more visual representation of inventory consumption and highlights emerging problems in the central market.

The difficulties of this system are that it requires many kanban cards that must be brought back in a timely and reliable manner if the batch board is going to reflect an accurate status of inventory in the market. Additionally, it takes resolve on the part of schedulers and supervisors not to look ahead, see what might be required next, and then start building it in advance of what actually is needed. Looking ahead is always a double-edged sword because you may avoid running low on inventory in some instances but are likely to overproduce in many others.

**Controlling Production Upstream—
Keys to Success**

- Determine how you will install pull production upstream from the pacemaker.

- Consider carefully whether you need a fixed-time conveyance cycle or a fixed-quantity conveyance cycle.

- Standardize the role of the material handler.

- Organize inventory in a central market location and size the inventory carefully.

- Use withdrawal kanban to pull inventory from the market to final assembly.

- Use signal kanban, such as the triangle kanban, to control batch processes upstream from the market.

Part 5 Expanding the System

Part 5   Expanding the System

10. How will you expand your level pull system across the facility?

Part 5: Expanding the System

Apogee had now converted their exterior-mirror value stream to a level pull system. From customer demand through assembly, painting, and molding, the pace was being driven by takt time with material replenished just in time.

Information and Material Flow for Exterior Mirrors

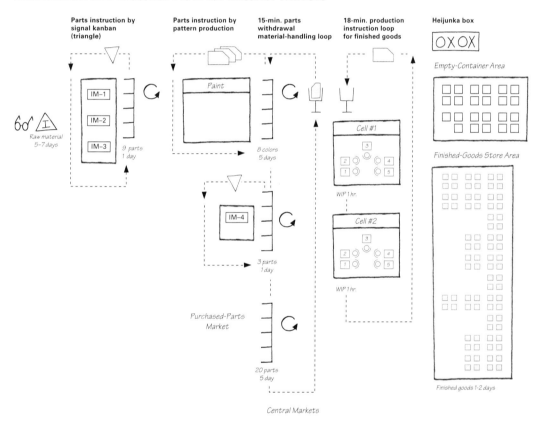

Multiple benefits to the value stream now were evident. On-time delivery to shipping and on-time delivery to customers were stable at 100%. The assembly cells were ready to assemble product 98% of the time—only isolated instances of minor downtime in upstream processes had prevented the cells from having a perfect week. For the first time in the collective memory of the improvement team, all the equipment in the value stream had run without overtime, and no expediting of material had occurred during the week. Dramatic increases in available space and dramatic reductions in inventory also were apparent: inventory had been reduced by 70% for work-in-process and 70% for finished goods since the start of the pilot.

Box Score—Exterior Mirrors

	Start of level pull	Current state	Comment
Productivity			
Direct labor (pieces per person per hr.)	11.0	12.5	14% improvement – reduced overtime
Material handlers	4	2	50% improvement – but low utilization
Quality			
Scrap	2%	1.5%	Rapid discovery of downstream problems
Rework	15%	12%	Rapid discovery of downstream problems
External (ppm)	105	105	No change – not addressed
Downtime			
Assembly (min. per shift)	20 min.	10 min.	Reduced waiting for material
Paint (min. per shift)	15 min.	15 min.	Reduced waiting for material
Molding (min. per shift)	10 min.	10 min.	Reduced waiting for material
Inventory turns			
Total	12	24	Estimated 2 weeks total on hand
On-time delivery			
To assembly	75%	98%	With no overtime or expedites
To shipping	85%	100%	With no overtime or expedites
To customer	100%	100%	With no overtime or expedites
Door-to-door lead time			
Processing time (min.)	125.7	125.7	No change
Production lead time (days)	30	12	Order throughput time of 2 days
Costs			
Overtime costs per week	$5,000	$0	100% reduction
Expedite costs per week	$2,000	$0	100% reduction

Productivity benefits now were beginning to surface as well due to level production and adherence to a consistent takt time. With overtime eliminated, direct labor was averaging a 13.6% productivity improvement (pieces per person per hour) for the most recent week.

Small gains in quality, although not a performance area specifically targeted during the implementation, were made as well. Compared to the past when it sometimes took several days or weeks for a defect to be detected, downstream problems were being surfaced much faster than before. Although the root causes were not solved by the level pull system, the symptoms were at least being detected much earlier resulting in fewer problems.

Indirect labor productivity, in the form of material handling, was improving as well because the exterior-mirror value stream now was serviced by a single conveyance operator who handled the production instruction loop and a single material handler running the parts withdrawal loop. This was despite their underutilization (30-70%) while serving only the exterior-mirror value stream, a problem Apogee will tackle next.

Question 10:
How will you expand your level pull system across the facility?

These improvements, although impressive, had been accomplished only on *one value stream* in the facility. The Apogee team now was at a transition point in their implementation and needed to think about expanding the improvements across all the value streams in the plant to capture the full benefits of system kaizen.

Apogee Current-State Layout

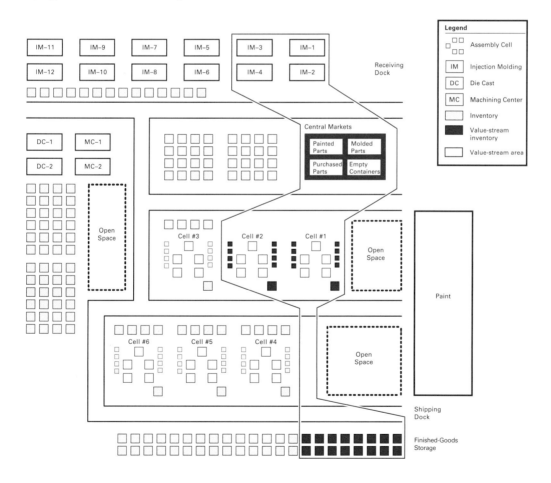

Project team members expressed varying opinions on how to move forward. One group wanted to continue to expand the system by value stream—a *value-stream rollout*. Since the exterior-mirror value stream improvement effort had taken approximately eight weeks to date and much of the elapsed time was learning the logic of the new system, they assumed the remaining value streams could be accomplished in no more than four months (for a maximum implementation time of six months).

Other members of the team instead proposed a *departmental rollout*, converting all the batch processes, then attacking all the remaining assembly cells and finished-goods areas at once, and finally targeting the central markets. Both approaches have their merits, as summarized in the table below.

Apogee Decision—Expansion Options

Option	Advantages	Disadvantages	Comments
Value-stream approach	• Pattern already developed • Implementation timing already known • Proven implementation results	• Five more cells • Lengthy rollout process • Difficult to partially schedule shared assets and may cause much confusion during transition period	• Not right at this time
Departmental approach	• Solves scheduling problem at shared assets • Quicker resolution of main problem areas • Capture of cross-value-stream efficiencies such as material handling	• Revised plan needed • Timing uncertain • Less experience	• Best fit for Apogee's current situation

Apogee realized that the value-stream approach normally worked best when all the assets in the plant could be neatly divided and dedicated to individual value streams: decisions and action items were clear because they did not affect other parts of the plant.

Unfortunately, many of the processes at Apogee used shared resources. All value streams went through paint. Die cast supported two value streams. Many molding machines were shared. Only the assembly cells and the four molding machines for exterior mirrors were completely dedicated to specific value streams. Scheduling shared assets always had been at the root of difficulties when push-scheduling the entire plant, and converting one area to level pull would set off a chain reaction that had implications for other areas.

As a consequence, Apogee decided to approach the rollout departmentally. The team would address all of molding together, all of paint and die-cast together, all purchased-parts together, and all assembly cells together—in that order. Apogee chose this approach because it would:

- Prevent some machines, such as molding, from being pushed on some part numbers while being pulled on others.
- Allow the team to immediately address the big problem area of supplying material from molding and paint to assembly and quickly help improve on-time delivery to customers.
- Make it possible to combine material handling across cells and value streams, capturing full benefits more easily.

Expansion Approaches

How should you proceed with the systematic rollout of your level pull system?

If your production assets are fairly light in nature, such as basic assembly operations, and dedicated to specific value streams, it is fine to implement the overall system one value stream at a time. As long as you can parcel out your factory into these easy-to-manage discrete pieces and work on one area while not affecting others, then it is logical to pursue the value-stream approach.

Unfortunately, most plants have shared assets, and it is tough to put only part of a batch machine on pull while other part numbers for other value streams run on push schedules. If you have shared assets and many batch-producing pieces of equipment, you will do better to take a whole "fix the department" approach. It may sound like more work, but the nature of the task is the same and it's easier to tackle it all at once. On the positive side, benefits will accrue faster and there is less chance that you are fixing one piece of the puzzle while causing damage elsewhere.

In either case, you eventually will have to think across all departments and all assets. Like Apogee, you will find that many tasks, such as material handling and scheduling, must be shared across the facility to maximize their efficiency. This is a subtle feature of *system kaizen* that is very powerful and can capture benefits that a vertical value-stream rollout initially will miss.

In the end, as always, weigh your needs against your capabilities and resources, and pick what works best for your specific situation.

To move forward with the departmental approach, Apogee held a meeting of the improvement team and the entire management. This group created a future-state layout for the entire facility. The team realized that it would be easier to set each piece in place once they had a firm vision of where all the pieces would go.

Future-State Overall Layout Concept

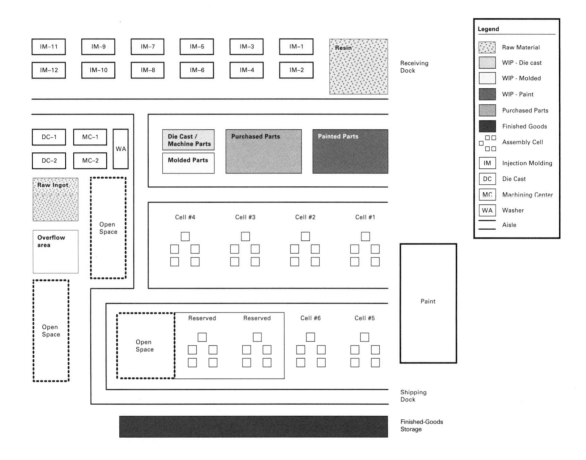

Based on this vision, Apogee's improvement team decided to implement four large rollout projects in the following order:

Step 1: Complete the signal kanban implementation for batch processes.

Step 2: Organize the remaining inventory into central markets.

Step 3: Establish a plantwide parts withdrawal kanban loop.

Step 4: Establish a plantwide production instruction kanban loop.

Step 1:
Complete the signal kanban implementation for batch processes.

Apogee chose to first complete the implementation of the triangle kanban in all the injection molding and die-cast areas. The team had learned from the initial pilot that upstream batch processes often were the problem in terms of getting the right part to the right place at the right time. Curing the big problems here first would have tremendous spillover benefits for the entire facility and the remaining value streams.

The process was straightforward and identical to their work with the exterior-mirror value stream (the methods described in Part 4). Machine by machine and part number by part number, the improvement team calculated the available time, the run time, the lot sizes, and trigger points for each part number in regular production.

Triangle kanban were created, inventory quantities were determined, and inventories for molded and cast parts were moved to a central market. On average there were three part numbers per molding machine in the plant. With 12 injection-molding machines, the molded parts section of the market held 36 part numbers with an average of one day of inventory.

Die-cast processes were similar, but with a slight twist. After the parts were cast they needed a small hole drilled and tapped. In this case, material could flow directly from the die-cast machine to a special machining center and washer unit on a first-in/first-out (FIFO) basis. There was no need to create a market between the two process steps. Instead a FIFO lane was created and a standard amount of work-in-process was established: two totes, or about one hour of material, between the processes to absorb minor downtime and fluctuation. The finished items then were stored in the central market.

Over a four-week period, Apogee was able to convert the entire molding and die-cast portion of the facility (approximately 10% of the parts) to pull scheduling with triangle kanban as the control mechanism. After some time trials it was found that one person needed to be dedicated to moving produced items from the machines to the market area for all of these machines. While in the market area, the same person was responsible for scanning the markets for inventory levels that had hit the trigger point for replenishment.

Signal Kanban Special Conditions

Keep the following points in mind when using signal kanban:

Signals for raw material: You will need to develop either a time-based or quantity-based method to signal the need for raw material for machines using the signal kanban. Otherwise, you'll lose valuable production time getting raw materials to the machines in time for product changeovers.

Nonproduction parts: You may need to run nonproduction parts in conjunction with the signal kanban replenishment system (e.g., engineering samples or infrequently run service parts that are not stored in inventory). For these items, you can make a generic *temporary* kanban that is painted a different color to call attention to the fact it will run only one time. If you need to run it periodically, you'll need to adjust your trigger points for replenishment of other parts made on that machine because the response times will be longer.

When this occurred, the individual took the triangle kanban and the appropriate amount of empty containers from the market to the designated machine. This cycle of activity dramatically lowered inventory across the plant and improved the linkage between the molding departments and downstream processes.

Step 2:
Organize the remaining inventory into central markets.

The team now moved to establish markets for all remaining parts, including those coming from suppliers (purchased parts).

Because lot sizes and trigger points for all the part numbers in injection molding and die cast had already been defined, the two remaining challenges for this step of the rollout were the interfaces with the paint department and purchase-parts storage area. The paint department was governed by the weekly production pattern (described at the end of Part 4) that dictated the order of the parts through the paint system and, hence, the replenishment time. The team merely had to identify all the regular part numbers and calculate the average demand for each of them over the past several months. This average demand multiplied by the one-week replenishment cycle and the addition of a small safety factor to cover for possible rework losses determined the amount of inventory in the central market for all 50 painted part numbers.

Each day at the shift's start, Production Control checked the inventory of painted items in the central market to see how it compared to target levels and, consequently, the amount to run that day. Production Control then issued basic quantity information to the paint supervisor on how much to run of each color. Because of the need to paint nonregular items—such as samples, engineering requests, and rare spare-part orders—the paint team reserved a small amount of open time at the end of every shift for this type of work. The open time also was used for preventive maintenance.

For the purchased-parts section of the market, the task was similarly straightforward. For each purchased part, a review was conducted to determine the appropriate replenishment and delivery interval for external suppliers. Approximately 35% of parts were from local suppliers that would deliver on a daily basis provided they received one-day notice on the exact quantity. An additional 50% came from in-state sources that would deliver only on a weekly basis. The final 15% of the suppliers were located far out of the region and some were in other countries. These suppliers shipped in bulk quantities on either a two-week or monthly delivery schedule. On average, however, about five days of inventory was held for each purchased part in the plant.

Replenishment of the market was conducted by Production Control using a simple check-and-order process. At predefined trigger points (e.g., the two-day level for weekly items) a Production Control analyst reviewed the items that were due to be ordered and fine-tuned the actual order amount as necessary, adjusting for scrap losses and other problems. Although not a pure kanban-replenishment system with the supplier, this basic method worked well enough for monitoring and controlling inventory at this point in Apogee's lean transformation. (For detailed instructions on setting up purchased-parts markets and calculating replenishment cycles with suppliers, see the LEI workbook *Making Materials Flow*.)

While establishing the central market the improvement team also came up with ideas to handle materials more efficiently.

- Flow racks for parts were marked and color-coded with signage to show when areas were full, at what point to reorder, and at what point—dangerously low levels—to expedite.
- The market was addressed by aisle, row, rack, and shelf, and this information was placed on the parts withdrawal kanban for each part.
- Heavy parts were put in easy-to-access locations.
- Frequently used parts were stored in locations near the end of the aisle where they were easy to pick.
- Parts of similar description were stored together for ease of locating by the material handler (storing parts by supplier was tried initially, but deemed less helpful).
- Signs and aisles were posted to show how traffic should flow through the markets.

Step 3:
Establish a plantwide parts withdrawal loop.

The current parts withdrawal loop in the plant covered only the exterior-mirror final assembly cell. To expand the withdrawal loop across all six cells, Apogee needed to first organize and address all of the cells and flow racks for ease of delivery. The improvement team would then create parts withdrawal kanban for all items at the assembly cells, and determine the right pitch for the withdrawal loop and the resources needed to run the loop.

For the parts withdrawal loop to function effectively, the addressing scheme was rolled out across all of assembly. Since there were only six final assembly cells and all were located in the same area, the simple method used earlier for addressing the exterior-mirror cells was continued (*see page 50*).

Each item that needed to be stored at assembly cells also needed a parts withdrawal kanban established. The improvement team created these kanban and filled in the needed information, including the address locations for pick up and delivery of product.

The team then had to determine the *total* number of material handlers to run the route, pick up the kanban, and deliver the parts.

The team estimated that across the six assembly cells, an average of 180 cards per hour would be triggered. The team earlier had estimated that a material handler could handle 10 to 12 cards on a 15-minute route, which implied that just over four material handlers should be needed to handle all the cells. This number was much less than the 10 material handlers previously used for each shift—and that was just for assembly-related material handling!

The challenge was how to best organize material handling for total efficiency. Dedicating material handlers to zones made the workload hard to balance because some areas triggered more parts than others due to varying pack quantities and takt times. Dedicating material handlers to individual cells could clarify responsibility, but this approach would require six material handlers, meaning that each would be underutilized.

Finally Apogee struck upon a winning solution: To balance work and total efficiency, Apogee would have all four material handlers serve all the cells in an integrated fashion. The team would work in a staggered, clockwise fashion. Imagine the overall parts withdrawal loop as a circle; the material handlers would rotate around the loop 90 degrees apart from each other.

After several attempts, this method worked well (*see graphic below*). Each material handler consistently picked up 10 to 12 kanban on what was now a full 15-minute route across all cells. The variable work component that changed in the equation was mainly the additional walk time to circle all six assembly cells for pick up and delivery. The rest of the work elements remained the same.

Four-Person Material-Handling Team

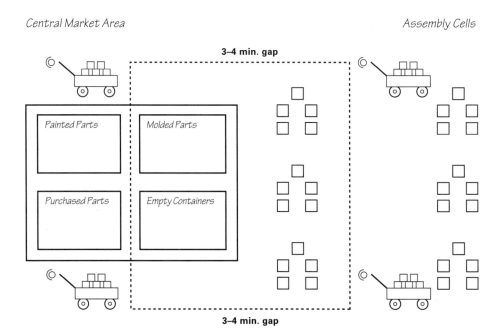

Updated Work Flow and Times for Withdrawal Material Handling

Step	Standard work elements	Conveyance time estimate
	Cell operator withdraws kanban card from the container upon removing the first part out of the container. The withdrawal kanban cards collect inside a kanban post at the station.	N/A
1.	On a fixed-time route the material handler travels from the central market and arrives at the assembly cells.	2 min.
2.	The material handler visits and picks up all the withdrawal kanban at each of the assembly cells as well as any empty containers, delivers parts to the cells, and returns to the market.	6 min.
3.	Once back at the store, the material handler sorts the cards for the best pick order.	15 sec.
4.	Empty containers picked up on the route are dropped off in the correct location in the stores.	1 min.
5.	Any necessary bins for small items, such as nuts, bolts, screws, washers, etc., are obtained.	1 min.
6.	Parts are picked from the store and placed on the material-handling cart (assume average of 12 withdrawal items each at 20 sec. per pick).	4 min.
7.	The withdrawal kanban are placed into the container for the appropriate item.	15 sec.
	Total time observed for plantwide withdrawal kanban rollout	14 min. 30 sec.

Once Apogee was able to standardize the system, there were several benefits to using a conveyance team:

- Multiple people were trained in the material-handling job and others could quickly be trained as well. The risk of lost skills due to retirement, vacation, or sickness was reduced.

- The facility now had a standard for determining material-handling work in assembly. When additional cells or parts came online the team could determine exactly how much extra work or resources this would involve.

- The focus for kaizen became clearer. The long walk time involved with the carts and the current route were not the optimal solution. Incremental improvement of 25% would allow the team to reduce the number of material handlers in the loop from four to three. If material handlers had been dedicated to cells, cross-value-stream improvements could not lead to gains in productivity.

Step 4:
Establish a plantwide production instruction kanban loop.

Much like the initial parts withdrawal loop in the plant, the production instruction loop currently covered only the two exterior-mirror cells in the facility. The current loop pitch for the pilot cell was still 18 minutes, with only 10 minutes of actual work for the material handler. To expand the instruction kanban loop across all cells, Apogee needed to determine the total workload for the production instruction kanban loop as well as establish the overall pitch for the withdrawal loop and the resources necessary to run the loop.

To determine the workload for the production instruction kanban loop, Apogee's improvement team reviewed the pack quantity and takt times for each of the six assembly cells in the plant as well as their respective pitch intervals.

Cell Workload for Instruction Kanban Loop

Area	Takt time	Finished-goods pack quantity	Pitch interval
Cell #1	54 sec.	10	540 sec. (9 min.)
Cell #2	54 sec.	10	540 sec. (9 min.)
Cell #3	60 sec.	12	720 sec. (12 min.)
Cell #4	60 sec.	12	720 sec. (12 min.)
Cell #5	45 sec.	8	360 sec. (6 min.)
Cell #6	45 sec.	8	360 sec. (6 min.)

As with most plants, there was no common pitch interval across the facility. Each product family had its own takt time and pack quantity, and Apogee needed to develop a loop to accommodate these differences.

Apogee wanted to follow its previous practice of creating standardized work and balancing the material handler's workload to the pitch intervals. Fortunately the basic work sequence of delivering the instruction kanban and empty containers, picking up the full containers from the cells, and delivering the product to the finished-goods area was standard across the cells.

Apogee calculated the total number of instruction kanban needed per hour per cell by dividing a typical hour by the pitch interval.

Instruction Kanban per Hour per Cell

Area	Time	÷	Pitch interval	=	Instruction kanban per hour
Cell #1	60 min.	÷	9 min.	=	6.7
Cell #2	60 min.	÷	9 min.	=	6.7
Cell #3	60 min.	÷	12 min.	=	5
Cell #4	60 min.	÷	12 min.	=	5
Cell #5	60 min.	÷	6 min.	=	10
Cell #6	60 min.	÷	6 min.	=	10
Total					43.4

Like the parts withdrawal kanban loop, Apogee's improvement team realized that much of the work in the instruction kanban delivery loop was fixed (e.g., walk time) and that the variable portion was directly related to how many containers were transported per trip by the person handling the material.

Regardless of how many containers were picked up, the total transportation time using a material-handling cart for finished goods was measured at roughly three minutes. Each finished-goods item picked up added about one minute of additional work. The team created several options for finished-goods material handling based upon this insight.

Apogee Decision—Options for Finished-Goods Material Handling

Potential Options	Fixed Time	Trips per hour	Kanban per trip*	Variable time**	Total time	Comment
18-min. loop	3 min.	3.3	13.2	13.2 min.	16.2 min.	OK. Current pilot pitch
15-min. loop	3 min.	4.0	10.9	10.9 min.	13.9 min.	OK. Even 4 trips/hr.
12-min. loop	3 min.	5.0	8.7	8.7 min.	11.7 min.	OK. Even 5 trips/hr.
9-min. loop	3 min.	6.7	6.5	6.5 min.	9.5 min	Overburdened
6-min. loop	3 min.	10.0	4.3	4.3 min.	7.3 min	Overburdened

* 43.4 kanban per hr. ÷ trips per hr.
** 1 minute incremental work per kanban

Apogee knew that the shorter the loop the more frequent the feedback to the cell regarding production status. However, short loops required more skill and the ability to respond quickly to problems in collaboration with line supervision. Having finally demonstrated the ability to maintain production to the 18-minute pitch in the trial phase, the Apogee team was ready to move to a shorter pitch interval.

Apogee realized that pitch intervals of either six or nine minutes were not attainable. Both of these options were rejected as next steps, but in the future they would be considered after additional kaizen.

Both the 12-minute loop and the 15-minute loop remained as options. With a 12-minute loop, instruction cards would be picked up five times per hour and approximately nine containers per trip would be handled. Although this was an aggressive target, with just 18 seconds of slack between each run, the conveyance supervisor felt confident of handling this interval with current methods and equipment.

Apogee decided to make the 12-minute loop the plantwide instruction kanban pitch. But there was one problem. The pitch interval for cells #1 and cell #2 would not match evenly with the 12-minute overall material-handling interval (e.g., nine minutes does not evenly divide into 12). Without some alteration there would be times when cells #1 and #2 would be left without an instruction kanban and, hence, the signal to build.

Pitch Interval Delivery

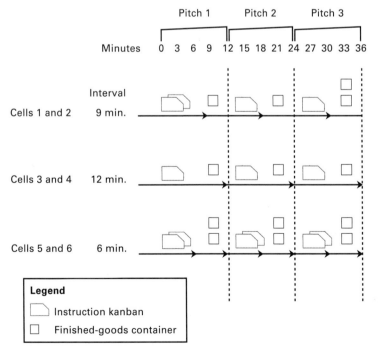

Instructions for Material Handler

Drop off 2 kanban at start of interval. Pick up 1 finished-goods container at end of interval and drop off 1 more kanban. Repeat next interval. Pick up 2 finished-goods containers at end of next interval and drop off 2 kanban. Cycle repeats.

Drop off 1 kanban at start of interval. Pick up 1 finished-goods container at end of interval and drop off next kanban. Cycle repeats.

Drop off 2 kanban at start of interval. Pick up 2 finished-goods containers at end of interval and drop off next 2 kanban. Cycle repeats.

To deal with cell #1 and cell #2, Apogee decided to stagger the delivery of information at these cells (*see Pitch Interval Delivery*). Instead of dropping off only one production instruction kanban at cells #1 and #2, the conveyance operator dropped off two instruction kanban on the first running of the route. This kept the cells running with 18 minutes' worth of production, until the route operator came around again 12 minutes later.

On the second trip, the conveyance operator dropped off one kanban card and picked up only one finished-goods container. This same basic pattern repeated on the third cycle, except that there were two finished-goods containers to pick up. The cycle then repeated. Because of their pitch interval and finished-goods container size, staggering was not necessary for the other assembly cells.

Dealing with Unmatched Pitch Intervals

Frequently you will find that pitch intervals do not align across cells. One option is to determine the lowest common integer applicable to each pitch. For Apogee, six, nine, and 12 all evenly divide into 36. Thus every 36 minutes cells #1 and #2 get four instruction kanban each; cells #3 and #4 get three instruction kanban each; and cells #5 and #6 get six instruction kanban each.

For some facilities in the initial stages of level scheduling and pull systems this option is fine. In Apogee's case, this interval was too long to pace the system and signal problems.

A second option is to dedicate material handling by value stream and product-family pitch interval. In this case, one material handler would handle cells #1 and #2, a second person would handle cells #3 and #4, and a third person would handle cells #5 and #6. If the workload for the material handlers matches the pitch interval, then this is an acceptable solution. However, this option also did not match up well for Apogee.

In some situations, finished goods can be moved away piece by piece from the cells and put in a shipping area for combined packing efficiency. But this is risky because parts might be mixed into the wrong pack containers, and proper steps must be taken to avoid errors. Other alternatives exist, such as changing the pack quantity of containers, but this, too, is risky. Takt time can change and disrupt your efforts to make pitch intervals match.

When the pitch intervals simply don't align, as in Apogee's case, you can always have some cells that get instructions slightly ahead of time to ensure that they will keep building, and then pick up finished goods on the next or following trip. This is what Apogee did for cells #1 and #2.

Expansion Complete

Apogee completed the rollout of the four major expansion steps over a four-month period as outlined. During this period it was necessary to build up some inventories and to make minor layout changes, which required some overtime and weekend work.

The information and material-flow map (*below*) for cells #1 through #6 shows Apogee's completed *system kaizen* to create a level pull system.

Information and Material Flow for Entire Facility

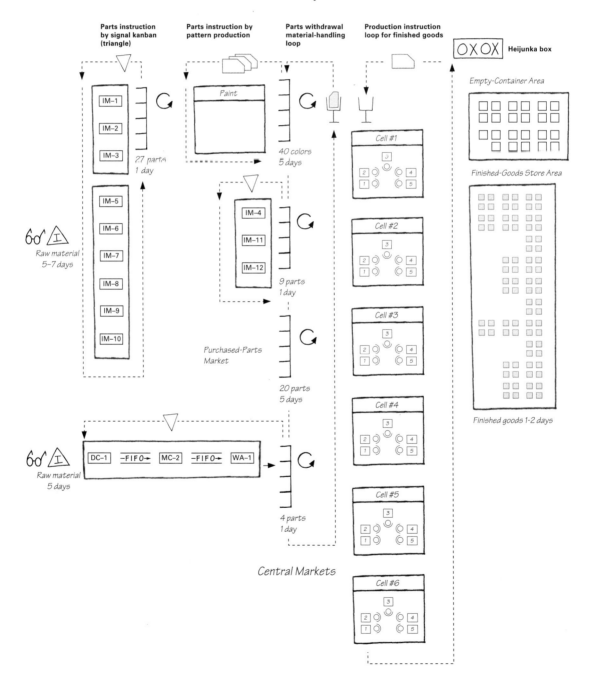

Expanding the System—Keys to Success

- Expansion via value-stream rollouts across the facility works best when equipment is dedicated and few assets are shared.

- Batch processes often need to be addressed collectively and not one process at a time.

- Capturing efficiency in scheduling and material-handling operations often means sharing resources across value streams.

- Establish plantwide parts withdrawal kanban loops and instruction kanban loops—do not establish individual loops.

- Sequence implementation steps carefully and tackle the biggest problem areas first.

Part 6 Sustaining and Improving

Part 6 Sustaining and Improving

11. How will you sustain your level pull system?
12. How will you improve your level pull system?

Part 6: Sustaining and Improving

The benefits of Apogee's plantwide step-by-step implementation of a level pull system were far-reaching, with improvements on many dimensions. Nothing changed with regard to the actual processing activities—molding, paint, and assembly—yet direct labor productivity went up substantially and indirect labor productivity went up dramatically because information and material flows now support rather than thwart the workflow. At the same time, operating expenses for inventories, space, overtime, and expediting were slashed, in some cases to zero, and Apogee became much more responsive to changes in customer requirements, particularly for product mix.

Despite their success over a six-month period (*as seen on the next page*), Apogee's improvement team knew that the complete implementation of the level schedule and pull system was not the end of their challenge. This event simply marked the need to shift focus from the lean conversion to creating a management system that could sustain and improve the new system over time.

Box Score—All Value Streams

	Start of level pull	Target	Current state	Improvement	Comments
Productivity					
Direct labor (pieces per person per hr.)	10.2	12.5	12.5	22.5%	No overtime
Material handlers (all shifts)	25	12	15	40%	Full utilization
Quality					
Scrap	2%	<1%	1.5%	25%	Not addressed
Rework	15%	<5%	12.5%	16%	Rapid detection
External (ppm)	105	<50	105	0%	Not addressed
Downtime					
Assembly (min. per shift)	30	<5	10	66%	Reduced wait for material
Paint (min. per shift)	15	<10	10	33%	Reduced wait for material
Molding (min. per shift)	20	<10	10	50%	Reduced wait for material
Inventory turns					
Total	10	30	24	140%	
On-time delivery					
To assembly	75%	98%	98%	23 points	No overtime or expedites
To shipping	85%	100%	100%	15 points	No overtime or expedites
To customer	100%	100%	100%	0 points	No overtime or expedites
Door-to-door lead time					
Processing time (min.)	125.7	125.7	125.7	0%	No change
Production lead time (days)	30	12	12	60%	Order lead time of 2 days
Other					
Overtime costs per week	$25,000	$0	$0	100%	No overtime
Expedite costs per week	$9,000	$0	$0	100%	No expedites

Question 11:
How will you sustain your level pull system?

Sustaining a level pull system is not an easy or short-term activity. Apogee needed to change its fundamental management practices for Production Control and Operations.

In any level pull system three management activities are critical:

A. Continuous monitoring of customer demand
B. Continuous assessment of performance metrics and process stability
C. Daily supervision of production control and operational processes to ensure that standard work is being followed

Someone must perform each of these tasks, but every organization probably will have a different allocation of responsibilities. Apogee assigned responsibility for its ongoing activities as described here, but you will need to have an explicit discussion and make precise decisions about the best allocation of responsibilities in your organization.

A. Continuous monitoring of customer demand

Customer demand (which is then divided by the available production time and converted to takt time) is the critical foundation of any level pull system. The amount of finished goods to hold, the amount of goods in the central market, and the hours of production to schedule are interlinked and tied directly to a calculation of *average demand* over a given period.

As average demand changes, Apogee must react by adjusting inventory levels throughout the facility. Apogee therefore established a monthly demand review led by Production Control and also attended by all of the Operations departments. The review assesses demand for each product.

For example, in Apogee's case we saw how one small improvement dramatically affected the level of finished-goods inventory (*see Part 3*). Before the schedule was leveled, finished goods for part #14509 were replenished on a weekly cycle, and this required a total of 1,200 parts to be held in the finished goods market. After the schedule was leveled and the A items were put on a daily replenishment, the amount of finished-goods inventory required for this part number was significantly lowered to 240 parts (*see next page*). The reason was simply that one of the key drivers of finished-goods inventory (the lead time required to replenish that inventory) had changed from five days to one.

Cycle Stock and Buffer Stock Variables

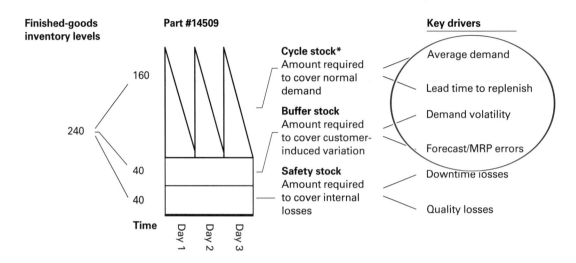

* Assumes an average customer demand and draw down during each day.

A change in other drivers also can change finished-goods inventory levels. For example, a reduction in average demand for this part number would further reduce the amount of cycle stock required. If daily demand falls from 160 items to 100 items, then the amount of finished-goods inventory needs to be reduced proportionally. Similarly, if downtime significantly increased, then inventory would need to increase in proportion.

A second dimension that should concern Apogee in addition to total demand for each product family and total inventory is *mix variations*. Often changes in mix are more important than changes in total demand, particularly in an operation such as Apogee's with a large number of paint colors and configurations.

For example, the paint market for the exterior mirrors holds five days' worth of inventory for each part number. The five days' worth of inventory is a relative number that depends upon average demand. Total demand for the value stream might easily stay constant, but the mix between two colors in the market might shift up by 50% on one item and down by 50% on another. The inventory in the market needs to accommodate this change. Otherwise when the parts withdrawal material handler goes to pick product for the assembly cell there will be too few of the items needed.

The final but critical component of Apogee's monthly demand review concerns *demand volatility*. Although Apogee has gone to great lengths to level demand within its walls (to the great benefit of both Apogee operations and its suppliers), there is no reason to think Apogee's customers have leveled their orders. Indeed, these may be getting more volatile even though their *average* demand remains constant.

As volatility of customer demand changes, Apogee needs to change the amount of finished goods held in its buffer inventory to protect the facility from the erratic demand waves. Doing this is a great use for statistical analysis with standard deviation calculations. For example, two standard deviations of demand for part #14509 equaled roughly 25% of demand or 40 parts daily or 200 parts weekly. An increase in volatility leads to a proportional increase in buffer stock. Apogee will need to adjust the amount of this product held in finished goods even though average daily demand is unchanged at 160. If Apogee managers fail to adjust finished goods in a period of rising volatility, waves of expedite orders will flood through the facility and risk capsizing the ship they have carefully transformed. And if volatility falls without anyone noticing, the facility can carry unneeded finished-goods inventory indefinitely.

The key questions related to customer-demand information for Apogee management to consider on a monthly basis include the following:

- Has average demand, mix, or volatility changed for any reason? If so, how will this affect:
    - Finished-goods inventory (cycle, buffer, and safety)?
    - Layout of finished-goods stores or shipping lanes or markets?
    - Inventory in markets?
    - Trigger points for batch processes?

- Has takt time changed for final assembly? If so, how will this affect:
    - Staffing levels in each assembly cell?
    - Pitch interval for instruction kanban?
    - Number of withdrawal kanban per hour triggered and required number of material handlers?
    - Withdrawal and instruction kanban loops?

B. Continuous assessment of performance metrics and process stability
At the individual manufacturing process level (assembly, paint, and molding) and in production control, it is important for Apogee to measure capability and performance over time. Without measurement, performance will backslide and the causes of the slippage will be a mystery.

At each process step it is essential to measure basic items such as scrap rates, changeover times, and downtime. Knowing the precise performance in each activity and charting it over time will show what improvement is needed. Performance is not just important for direct operations but also for inventory levels because the safety stocks in the system are directly related to the likelihood that the right amount of products can be produced at the right time.

Failure to monitor performance will result in *ineffective* markets that fail to absorb process instabilities and protect the customer. However, as you calculate safety stocks please remember that no market can *solve* an instability problem. The only solution is targeted kaizen as problems are prioritized, root causes are determined, and permanent fixes are put in place.

Key questions related to process stability for Apogee to consider include:

- What metrics do you need in the value stream to monitor the stability of performance (on-time delivery, scrap, rework, productivity, safety, downtime, etc.)?
- How often will you measure and update these items for the value stream?
- Who is responsible for tracking, reporting, and improving these items?
- Where exactly is the instability in each value stream (e.g., paint, molding, die cast)?
- How large is the instability and what amount of inventory and lead time is it driving?
- What point kaizen events will you need to solve these problems?
- What do you need to improve first?

C. Daily supervision of production control and operational processes to ensure that standard work is being followed

The final activity necessary for Apogee to sustain level pull scheduling is active supervision by Production Control and Operations. There is no scheduling system that works perfectly according to plan, day in and day out, so these two departments will need to make adjustments. The techniques described in this workbook are ways to set a plan in place, but they are intended to be flexible enough to enable you to react to the inevitable changes that occur.

In one sense, Apogee's level pull system is the ultimate tool for uncovering sets of problems that existed in the facility but were previously unnoticed—defects, overproduction, late deliveries from suppliers, lengthy changeover events, and material stored in wrong locations. However, it still is the role of Apogee supervisors to detect these systematic problems and make sure they are addressed and solved over time.

With that in mind, Apogee made anything important to the success of the overall system *highly visual*. This will increase the odds for sustaining improvement because everyone will catch abnormal conditions faster and react to them quicker. Talking with people or getting reports from subordinates is one way for supervisors to obtain information, but it is passive and may miss or filter out critical items. Active management is required among front-line supervisors for lean manufacturing to succeed.

In implementing a more active and visual style of management for each value stream, Apogee managers will need to seek answers daily or even more frequently to several questions:

- Is production ahead or behind schedule?
- Are inventory levels above or below normal?
- Do we have the right number of resources in place?
- Are machines producing to cycle time?
- Is assembly producing to takt time?
- Are defects occurring and escaping downstream?
- Are suppliers delivering on time?

Making the answers to these questions—which seek to identify abnormal conditions—obvious and visual will go a long way toward helping Apogee sustain the gains from its level pull implementation.

Question 12:
How will you improve your level pull system?

By applying standard management to its processes, Apogee was able to sustain the gains it had made in its level pull system. The next and ongoing challenge for Apogee and for every facility is to improve the system.

In Apogee's case there still is a considerable amount to improve. By design, several short-term compromises and trade-offs were made during the implementation that will gradually need to be resolved. This is a typical pattern and almost certainly will apply to your facility.

Process Stability

For each of the production processes in Apogee, there is some remaining waste that manifests itself in the form of availability, speed, or quality losses. All of these types of losses must be compensated for in your inventory calculations in the form of safety stock. Once safety stock is created, however, it does not mean that Apogee has license to leave it there forever. Unlike customer demand and demand variation, which is out of Apogee's control, safety stock can be reduced over time as causes of instability are eliminated.

Stability Improvement Levers

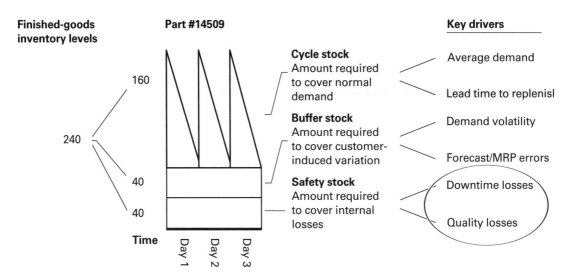

At each step of the system, Apogee still has 10 to 20 minutes per shift of downtime and 2% to 15% quality instability in the form of rework and scrap. These problems all result in safety-stock inventory (central markets and finished goods) that must be held to buffer the system against internal sources of volatility. For part #14509, approximately 17% of the total inventory is being held to buffer against these internal losses, and this is typical.

Apogee will need to target and reduce safety-stock inventory through a series of targeted kaizen efforts. Apogee should focus on:

- Reducing scrap and rework in the paint department, specifically the problem of inclusions (small particles adhering to the paint).

- Eliminating equipment downtime in molding, paint, and final assembly, specifically minor mechanical downtime related to limit switches and sensors.

- Eliminating delays caused by material-handling problems between processes, specifically between the producing departments and the central markets as well as between the central markets and final assembly.

Other point kaizen activities, such as set-up and changeover reduction, are important as well. Reducing changeover time will shorten the lead time to replenish cycles and help reduce cycle stock, which subsequently reduces a fraction of the safety stock as well.

Customer-Induced Variation

Another important objective is decreasing inventory buffers set in place because of customer-induced variation. Ultimately the root cause of customer variation may be entirely out of manufacturing's control, but a closer look always is warranted.

In Apogee's case there were a few part numbers in the door-handle value stream that suffered tremendous amounts of volatility. For example, while the average demand for molded part #13901 was 1,000 units per week, the variation was up to five times this amount in a three-month period.

Investigating this specific case revealed that the part number in question was used by a sister plant located two hours away. Since the demand for this item was fairly low, the customer-service department at headquarters held orders for several weeks before releasing them to the Apogee mirror plant. While demand was fairly constant, the practice of *lumping* orders created demand spikes and volatility that did not really exist.

When confronting high variation, you will need to peel back the onion several layers to find the root cause. Often, in larger companies where the customer-service and order-entry functions are located external to the plant, information is not shared in a timely fashion.

Other customer-induced variation problems actually may be caused by internal pricing decisions and sales behavior. Often customers are trained to expect end-of-the-quarter price breaks or special volume-purchase deals. These types of variation are outside of the control of plant operations. Operations can and should, though, question whether they provide value to the organization that offsets the variation and disruption they cause.

Systematic improvement potential

A major source of system improvement can be found in nonproduction time, specifically waits in inventory markets that constitute 99% of the nonvalue-creating time for products as they proceed through the facility.

Cycle stock can be reduced only by shortening the door-to-door lead time to manufacture. Apogee had done an excellent job of reducing the required order lead time for any one unit from several weeks to two days (the time from when the instruction kanban was sent to the assembly cell to the time of customer shipment), but total manufacturing lead time was still approximately two weeks for the longest lead time item (molding in this case). For overall system improvement, the greatest opportunity still lies from the *assembly cell backward to raw material*.

System Lead Time vs. Order Throughput Time

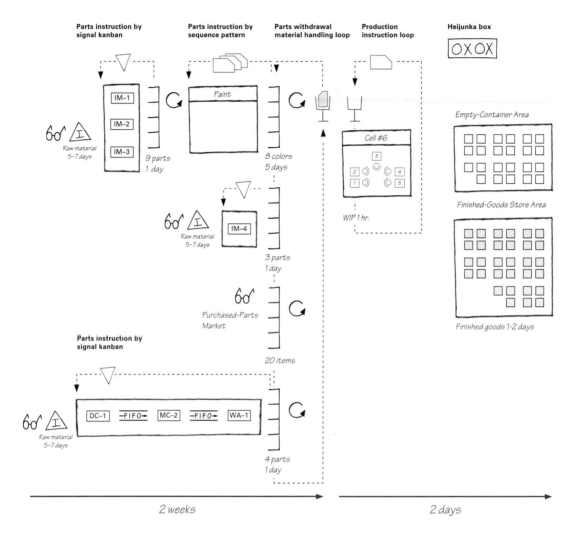

In the case of the longest lead-time item (painted injection moldings) there are three locations where inventory is kept upstream of the assembly cell: the market after paint, the market after molding, and the raw-materials storage area. Each of these can be examined more closely to find system improvement potential.

Improvements at Markets

Market	Quantity	Lead-time driver	Possibilities
Painted parts	5 days	1-week replenishment cycle	• Decrease paint batch size • Add second shift in paint • Send instruction from assembly to paint
Molded parts	1 day	1-hr. changeover time	• Reduce changeover time • Add third shift
Raw material	5 days	Supplier delivery • 50% weekly • 35% daily • 15% 2-4 weeks	• Increase frequency of supplier delivery of current supplier • Identify new local suppliers

1. Painted-parts market

The painted-parts market is driven by a one-week replenishment cycle of the paint department and the need to paint 50 colors across five production days. On average 10 colors are being painted per day during the one shift the paint department operates. In order to reduce the replenishment lead time there are three main options for Apogee to weigh in terms of cost and benefit.

The first option would be to reduce the batch size in paint and paint more colors per day. Increasing to 20 colors per day would equate to a 50% reduction in painted material in the market. Painting all 50 colors in one day would mean that replenishment time could be taken down to one day. The cost of extra paint purging would have to be weighed against the value of the reduced inventory, but it often is the case that clearly identifying a large opportunity inspires technologists to rethink a whole process.

The second option for paint, assuming other factors are constant, would be to decrease the line speed and run a second shift of paint production. This would cut the replenishment lead time in half and reduce the market inventory by 50% as well. Obviously the effects on quality and operating cost must be factored against the value of the reduced inventory.

If these options are feasible, especially the first, then Apogee should consider whether it needs the painted-parts market. If the paint process is stable enough and lead time to replenish is short enough, then production instruction kanban can be sent from the assembly cells directly to the paint department to signal replenishment (*see page 102*). However, the cells will need to increase the amount of painted inventory held at the line to cover any increase in replenishment time.

Direct Replenishment Instruction from Assembly to Paint

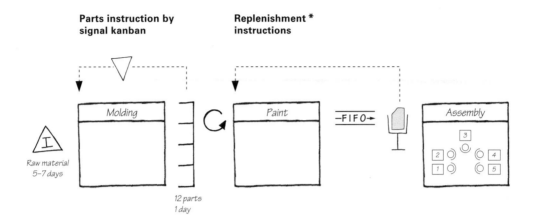

* In this instance, the withdrawal kanban functions as an instruction kanban for the paint department.

2. Molded parts market

The key lever for reducing inventory in the molding market is further reduction of the current changeover time of 55 minutes. Each minute that can be converted to production time reduces replenishment time. This in turn enables Apogee to lower lot sizes and trigger points for replenishment and reduce lead time.

In some situations and without capital investment, companies have been able to reduce set-up and changeover time by 80%. Over longer periods of time and with cooperation from the maintenance staff and equipment manufacturers, changeover times on some processes have been reduced to the single-minute level. If this can be achieved, the need for holding molded parts in markets should be questioned.

If changeover times cannot be reduced, Apogee could increase the time available for production by adding more shifts and using this time to conduct more changeover events. The additional cost would need to be carefully measured against the benefits of reducing inventory from the already low level of one day.

3. Raw material

The last area for reducing lead time is in the raw materials portion of Apogee's value streams. These were not a significant problem for on-time delivery or quality, but because there is an average of five days' inventory on hand and a mix of supplier delivery frequencies, these elements of Apogee's value streams eventually will be a target for system improvement.

Apogee can investigate the possibility of more frequent delivery directly from suppliers, switch to local suppliers, and change the logistics pattern for pick up of supplier material. Of course, these options have to be evaluated in terms of cost, quality, and inventory reduction to determine the benefit to the overall system.

The objective in every case must be to look at benefits vs. costs for the total system rather than at individual areas or process steps.

> **Sustaining and Improving—Keys to Success**
>
> - Monitor changes in customer demand and make adjustments to inventory as needed.
> - Measure key process activities and monitor them for improvement.
> - Practice visual control—if it is important, make it visual.
> - Improve process stability over time.
> - Reduce changeover time on internal processes.
> - Investigate system design options that eliminate the need for markets between processes.

Conclusion

This workbook illustrates a company taking a major leap in lean manufacturing by implementing a level and pull-based production-control system across all value streams within an entire facility. While previous Lean Enterprise Institute workbooks have focused on point and flow kaizen, *Creating Level Pull* aims to help manufacturers tackle system kaizen to introduce lean production scheduling.

Implementing level and pull production across several value streams and shared production assets is a difficult task in any environment. In pursuing this type of improvement, you will face many decisions on which path to take *and a series of options*. There are no simple answers applying to every situation. Basic issues such as your current level of internal process stability, length of manufacturing lead time, and the nature and frequency of your customer orders (As, Bs, and Cs) will affect many of your decisions.

The Apogee example shows that a level pull system can have tremendous benefits for any company, especially one with discrete processing and repetitive demand. Often the step of implementing level and pull methods for scheduling is an important enabler for stabilizing the pace and consistency of production within a facility. Once level and pull systems are implemented the tasks of operating cells to takt time and making material flow between locations become much simpler. Apogee's case shows that the measurable benefits of a level pull implementation are large.

As you approach your implementation, I encourage you to focus on the following activities and thought processes:

- Treat your lean implementation as the systematic improvement of interconnected process steps across the value stream. Don't just fix isolated areas or apply random improvement tools. You'll have greater impact and sustainability if you structure your implementation as outlined in this workbook.

- Use value-stream mapping in conjunction with level and pull-based scheduling to highlight your next set of improvement opportunities. Don't merely assume the location and cause of problem areas. The implementation of level and pull production will aid you in seeing more clearly where waste, overburden, instability, and variation exist.

- Remember to think and improve *across* value streams as well as *within* them. Often there are huge improvement opportunities to be realized by attacking the wastes inherent in shared production assets that are batch in nature, as well as addressing shared tasks such as scheduling and material handling. Focusing on single value streams will miss large opportunities for improvement.

I wish you great success when implementing your level pull system, and I look forward to hearing about your progress.

About the Author

Art Smalley

Art Smalley learned about lean manufacturing while living, studying, and working in Japan for 10 years. Art was one of the first foreign national employees to work for Toyota Motor Corp. in Japan and spent the majority of his Toyota career from the late 1980s to the mid-1990s helping Toyota transfer its production, engineering, and management system to facilities around the world.

As part of his education at Toyota, Art was assigned to the Kamigo Engine Plant. This is Toyota's largest engine plant and the facility where Taiichi Ohno made many important discoveries while serving as the founding plant manager.

After leaving Toyota, Art served as director of lean production at Donnelly Corp., a global Tier One automotive supplier with plants in 15 countries. (Donnelly is now part of Magna Inc.) More recently, Art helped start the lean manufacturing practice of McKinsey & Co., where he is still an advisor. Art currently divides his professional time between aiding companies implementing lean systems in the U.S. and acting as managing director of SEI Inc., an engineering and maintenance services company in Japan.

Appendix: Kanban

Appendix: Kanban

Introduction to Kanban

In lean manufacturing, the kanban is the specific tool for controlling information and regulating materials conveyance between production processes. (The term is Japanese, meaning "sign" or "signboard.") Kanban coupled with takt time, flow processing, pull production, and level scheduling is what enables just-in-time production to be achieved in a value stream. Typically a kanban is used to signal when product is consumed by a downstream process. In the simplest case this event then generates a signal to replenish the product at the upstream process.

Kanban differs from traditional production control methods in several important ways. In traditional manufacturing, the production schedule is provided to each individual process and each process produces in accordance with that schedule (without timely feedback from downstream processes regarding exact needs). Kanban instead functions as a physical schedule tool that tightly links and synchronizes production activity between upstream and downstream processes. Furthermore, in traditional manufacturing the movement of material between processes occurs when the upstream process completes the product. This results in a push of material to the next station, independent of the exact needs of the downstream process. Kanban instead combines control over movement of material with respect to both time and quantity dependant upon signals from the downstream process. Thus, kanban controls production in a value stream by controlling material and information flow.

Traditionally within a single facility, kanban is a simple paper card, sometimes protected in a clear vinyl envelope. The card contains basic information such as part name, part number, external or internal supplying process, lot size, pack quantity, storage address, and consuming-process location. A bar code may be printed on the card for tracking or automatic invoicing. When communicating over long distances, electronic signals often are used in place of simple kanban cards.

Purpose of Kanban

There are four major purposes for kanban:

- Prevent overproduction (and overconveyance) of material between production processes.
- Provide specific production instructions between processes based upon replenishment principles. Kanban achieves this by governing both the timing of material movement and the quantity of material conveyed.
- Serve as a visual control tool for production supervisors to determine whether production is ahead or behind schedule. A quick look at the devices that hold kanban in the system (kanban accumulation posts) will show if material and information are flowing in timely accordance to plan or if abnormalities have occurred.

- Establish a tool for continuous improvement. Each kanban represents a container of inventory in the value stream. Over time, the planned reduction of the number of kanban in a system equates directly to a reduction in inventory and a proportional decrease in lead time to the customer.

Types of Kanban

There are two main types of kanban: *production instruction* kanban (also known as *make* kanban) and *parts withdrawal* kanban (also known as *move* kanban).

Types of Kanban

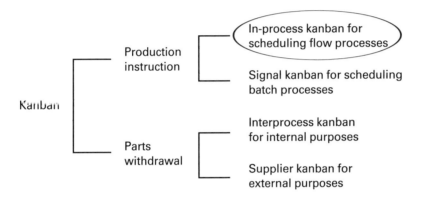

The distinguishing feature between the instruction kanban and the withdrawal kanban is that the former is a signal to make something, while the latter signals that something needs to be removed from inventory (which then triggers replenishment) and conveyed to a downstream process. Each type of kanban has two further subdivisions (*see Types of Kanban graphic*).

In-Process Kanban

The *in-process* kanban is used to convey make instructions for small amounts (ideally one-by-one production or at least one pitch corresponding to one pack quantity) to an upstream process. Typical uses include scheduling final production areas based upon withdrawal of inventory from a market or a direct replenishment signal from a customer. An example of the type used at Apogee is shown on the next page.

In-Process Kanban

```
Finished-goods          Product              Production
storage information   information area         area

Finished-goods       Part #:              Line address:
market                                    Assembly Cell #1
                     14509

Supplier:            Part description:
Internal             LH Exterior Mirror
                     Heated Black         Location:
                                          Final Assembly
Market location:     Quantity: 10         Department
A4
R2
```

Signal Kanban

The *signal* kanban is used to convey make instructions for large quantities to upstream batch processes such as stamping presses and molding machines. In-process kanban would be less effective in these applications due to the large number of cards required and the associated time to handle them. Instead, signal kanban utilizes lot size in conjunction with markets to feed downstream processes while still allowing time for changeover work to occur at upstream processes. There are three variations on the signal kanban: pattern production, lot-making, and triangle kanban.

Pattern production is an effective method for scheduling processes when there is an optimal order or sequence for production to follow because of either the types of materials in use or a changeover sequence that must be accommodated. For example, there often is an optimal changeover sequence when processing different gauges of steel, chemical formulas, paint colors, etc. In pattern production a basic sequence is established and adhered to, but the lot size produced each time may vary. In this fashion a fairly stable production lead time and time interval for producing every part can be established. The lead time for replenishment for the pattern is then used to set the inventory level in the market.

Another form of signal kanban, called *lot-making*, uses batch boards in conjunction with inventory markets. Each item in the market has a kanban that is detached and returned to the upstream producing process as inventory is consumed. Once the kanban cards accumulate to an established amount (trigger point), replenishment begins in accordance with the number of cards. This form of signal kanban differs from an in-process kanban in that cards are grouped into a production lot, rather than production occurring one card at a time.

The most widely used form of signal kanban is known as the *triangle kanban* (the name comes from its distinct triangular shape). Triangle kanban are used to schedule a batch process that has substantial changeover times and machine cycle time significantly faster than the takt time of production downstream. This kanban uses a lot size for production in conjunction with a trigger point to replenish inventory and is frequently used for stamping, injection molding, and similar processes. A key benefit of the triangle kanban is that only one kanban per part number is created—multiple cards do not need to be managed.

An exampled of the type used at Apogee to regulate the molding process is shown:

Triangle Kanban

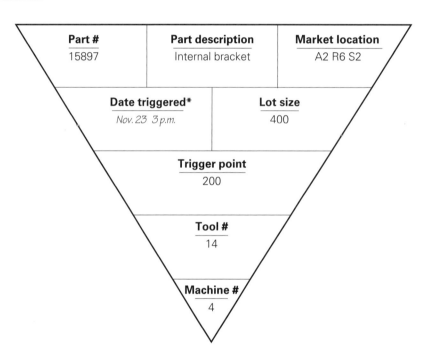

* The point in time at which the kanban is placed on the signal kanban rail. This aids production management in assessing status of the process and reduces the possibility of getting kanban out of sequence.

Interprocess Kanban

The *interprocess kanban* is used to signal the need to withdraw (move) parts from a storage area and convey them to a downstream process *within* a facility. This type of kanban normally is employed in conjunction with continuous-flow assembly cells that use a large number of components from either internal or external sources. A prerequisite for use of the withdrawal kanban is the creation of a material market as well as the determination of storage quantities at line side. The intent of this kanban is to enable the storage of small quantities of material at the final assembly area in order to maximize the space available for production. This in turn requires that the production assembly cell be supplied with frequent and regular delivery of small quantities of material. In order for this system to work, each location must have dedicated positions and addressing mechanisms in place for ease of material movement. An example of interprocess kanban used at Apogee to withdraw parts from the central market and convey them to the point of use in assembly cell #1 is shown:

Interprocess Withdrawal Kanban

Supplier information area	Product information area	Point-of-use area
Supplier code: ABC	**Part #:** X2174	**Line address:** Station #1 Flow Rack #2
Supplier: Ajax Springs	**Part description:** Spring	
		Line location: C1 S2 R2
Market location: A2 R7 S1	**Quantity:** 40	

Supplier Kanban

The *supplier* kanban is used to signal the need to withdraw parts from an external supplier for conveyance to a purchased-parts market or central market at the downstream customer. The supplier kanban differs from the interprocess kanban in that it is used with external suppliers.

In advanced applications the supplier kanban also contains information that is known as a *kanban cycle*. An example of a kanban cycle is the notation *1:4:2*, which indicates that for a particular part number the supplier in *one* day will deliver that part and pick up kanban *four* times, and the kanban picked up on any given cycle will be returned with the requested parts *two* trips later. In this case, since there are four deliveries each day (at six-hour intervals), the requested material will return in 12 hours. (Note that the supplier often is delivering other materials throughout a plant, adhering to additional kanban cycles, and delivering parts for some locations while picking up kanban for others throughout the day.) Various kanban cycles can be created depending upon the nature of the supplier's production process, the amount of supplier finished-goods inventory, and the supplier's distance from the customer.

An example of a supplier kanban used by Apogee to withdraw parts from Ajax Springs is shown:

Supplier Kanban

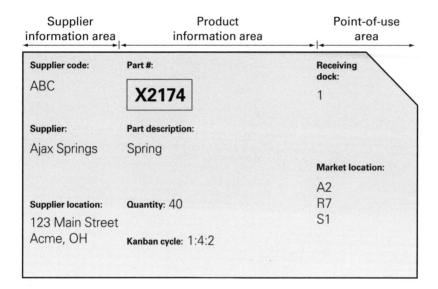

Temporary Kanban

Not depicted on the kanban classification chart is what is often referred to as the "temporary kanban." Typically the number of kanban cards are calculated and regulated monthly by Production Control, and the quantity of cards is only subject to change monthly based upon changes in demand and lead time. Often, however, there are short-term events that require additional kanban to be injected into the system for smooth production. Reasons to do this include the build up of inventory to adjust for the differences in working days between customer and suppliers, or to make up for time spent on die maintenance or machine repairs. The temporary kanban, as the name suggests, is for one-time use only and should clearly show an expiration date. It is good practice to color code or identify these kanban in some special manner so that they are not accidentally kept in the system after their intended use is completed.

Kanban in Combination

In general, production and withdrawal kanban are used in combination to control pull production between processes where some work-in-process must be stored in markets. This arrangement works at Apogee as shown for regulating pull production between final assembly, the central market, and paint:

Production and Withdrawal Kanban

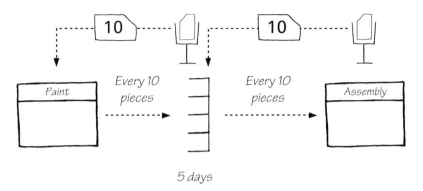

A similar use of production and withdrawal kanban in combination is shown among final assembly, the central market, and molding:

Signal and Withdrawal Kanban

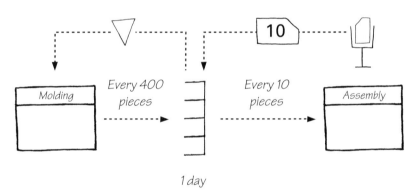

Finally, as a word of caution, kanban usage must be taken into consideration in conjunction with conveyance routes in the facility. There must be a standard conveyance mechanism in place if kanban cards are to control the flow of production. The principle of conveyance in lean manufacturing is to deliver mixed loads of products to downstream processes as requested in a timely, frequent manner. Normally a *fixed-time variable-quantity* style of conveyance is used within a facility or between facilities to supply product in conjunction with kanban. This requires that the conveyance route be on a precisely timed route with a repeatable standard pattern. In certain instances between processes, a *fixed-quantity variable-time* method also may be used to signal for certain type of material, particularly items that are used less frequently. In this latter case the key factor is to decide how to signal for the item requested and when the signal needs to be made in order to prevent operations from running out of material.

References

Harris, Rick, Chris Harris, and Earl Wilson. 2003. *Making Materials Flow.* Cambridge, MA: Lean Enterprise Institute, Inc.

Jones, Dan, and Jim Womack. 2011. *Seeing the Whole Value Stream.* Cambridge, MA: Lean Enterprise Institute, Inc.

Marchwinski, Chet, and John Shook, compilers. 2003. *Lean Lexicon.* Cambridge, MA: Lean Enterprise Institute, Inc.

Rother, Mike, and John Shook. 1998. *Learning To See.* Cambridge, MA: Lean Enterprise Institute, Inc.

Rother, Mike, and Rick Harris. 2001. *Creating Continuous Flow.* Cambridge, MA: Lean Enterprise Institute, Inc.